Spectroscopic Tools for Food Analysis

Spectroscopic Tools for Food Analysis

Edited by
Ashutosh Kumar Shukla
Ewing Christian College, Gaughat, Allahabad, India

IOP Publishing, Bristol, UK

ISBN 978-0-7503-2324-6 (ebook)
ISBN 978-0-7503-2322-2 (print)
ISBN 978-0-7503-2323-9 (mobi)

DOI 10.1088/2053-2563/ab4428

Version: 20191101

IOP ebooks
ISSN 2053-2563 (online)
ISSN 2054-7315 (print)

British Library Cataloguing-in-Publication Data: A catalogue record for this book is available from the British Library.

Published by IOP Publishing, wholly owned by The Institute of Physics, London

IOP Publishing, Temple Circus, Temple Way, Bristol, BS1 6HG, UK

US Office: IOP Publishing, Inc., 190 North Independence Mall West, Suite 601, Philadelphia, PA 19106, USA

To my parents

Contents

Preface ix

Foreword x

Editor biography xii

List of contributors xiii

1 State of the art and applications of near infrared spectroscopy in food composition and quality **1-1**

1.1 Introduction 1-1
1.2 Spectroscopic techniques 1-2
1.2.1 Near and mid infrared spectroscopy 1-2
1.2.2 Raman spectroscopy 1-3
1.2.3 UV and visible spectroscopy 1-3
1.2.4 Hyperspectral and multispectral imaging spectroscopy 1-5
1.3 Multivariate data analysis 1-6
1.4 Applications 1-8
1.5 Summary 1-9
References 1-10

2 Food authenticity and traceability: molecular spectroscopy approaches **2-1**

2.1 Introduction 2-1
2.2 Molecular spectroscopy 2-2
2.3 Chemometric methods 2-4
2.4 UV/vis spectroscopy 2-6
2.4.1 Food components, pigments, and contaminants 2-6
2.4.2 Food adulteration and fraud 2-7
2.5 Fluorescence spectroscopy 2-8
2.5.1 Food compositions, additives, and contaminants 2-9
2.5.2 Food authentication, origin, and adulteration 2-10
2.6 IR spectroscopy 2-12
2.6.1 NIR spectroscopy 2-15
2.6.2 MIR spectroscopy 2-18
2.7 Raman spectroscopy 2-21
2.7.1 Food components analysis 2-22
2.7.2 Food authenticity and adulterations 2-24

2.8 NMR spectroscopy 2-24
2.8.1 ^{1}H NMR 2-25
2.8.2 ^{13}C NMR 2-28
2.8.3 ^{31}P NMR 2-28
2.9 Conclusions 2-29
Acknowledgments 2-30
References 2-30

3 Nuclear magnetic resonance spectroscopy applications in the food industry 3-1

3.1 Nuclear magnetic resonance spectroscopy 3-1
3.2 Use of NMR techniques in food science 3-2
3.2.1 Application of NMR spectroscopy to milk and dairy products 3-3
3.2.2 Application of NMR spectroscopy to the beverage industry 3-5
3.2.3 Application of NMR spectroscopy to the food oil industry 3-6
3.2.4 Application of NMR spectroscopy to specific foods 3-7
3.2.5 Application of NMR spectroscopy to fruit and vegetables 3-8
3.2.6 Application of NMR spectroscopy to meat products 3-8
3.2.7 Application of NMR spectroscopy to food emulsions 3-9
3.3 Results 3-9
References 3-9

4 Spectroscopic method for the detection and determination of ammonia nitrogen in aquaculture water 4-1

4.1 Introduction 4-1
4.2 Spectroscopic method for the detection of ammonia nitrogen 4-3
4.2.1 Absorption spectrometry 4-3
4.3 Fluorescence spectrometry 4-6
4.3.1 O-phthalaldehyde (OPA) 4-6
4.3.2 Other methods 4-9
4.4 Conclusion and future perspectives 4-10
References 4-10

Preface

Quality of food items has been a concern throughout time. The need for suitable techniques for quality assessment has therefore attracted the attention of the scientific community. Different spectroscopic techniques have proven their importance and have been applied as the method of choice accordingly. This book is an effort to bring to the reader the different spectroscopic techniques and their applications in food analysis.

This volume covers infrared, Raman and NMR spectroscopic methods and includes the analysis of raw food materials, food emulsions, beverages, dairy products, fruit and vegetables. Aquatic plants and animals serve as animal protein sources in our diet and the source of protein synthesis in aquatic organisms is ammonia nitrogen. A chapter on spectroscopic methods applied for ammonia nitrogen detection in aquaculture water has therefore been included.

I sincerely thank Jessica Fricchione, Senior Commissioning Editor at IOP Publishing for giving me an opportunity to present this book to readers. I wish to thank Daniel Heatley, ebooks Editorial Assistant and Sarah Armstrong, Editorial Assistant, IOP Publishing for extending all the support during the development of this project. It is the prompt response of Poppy Emerson, Production Editor, IOP Publishing, which led me to present this work in a short time. I thank the expert authors for taking time out of their busy academic schedule to contribute to this volume.

Ashutosh Kumar Shukla
Prayagraj, India
September 2019

Foreword

The purpose of the edited volume *Spectroscopic Tools for Food Analysis* is to provide a distinctive multidisciplinary review of modern trends in spectroscopic analysis of food products. Today's spectroscopy methods are well developed and adequately positioned for analysis of matter at the molecular level and generally afford a wealth of information on chemical composition of matter and even molecular structure. Historically, spectroscopy originated from dispersing the visible light according to its wavelength by a prism and studying interaction of such a light with matter. Nowadays, spectroscopy encompasses a very broad range of electromagnetic radiation interacting with the sample. Combined with rapidly developing computers, electronics, and also electromagnetic radiation sources over a very broad frequency range—from radiofrequency and microwave devices for cell phone communication to inexpensive diode-based lasers—this technical progress leads to an increasing availability of inexpensive and even portable spectrometers that could be applied for the analysis of food and agricultural products.

The individual chapters assembled in this edited volume provide a representative sampling of the spectroscopic methods as applied to raw food ingredients and agricultural commodities as well as processed food products. The volume starts with a review of state-of-the-art methods of near (NIR) and mid infrared (MIR) spectroscopy as an analytical tool for the modern food industry. NIR and MIR are based on absorption of photons by molecules by exciting specific vibrational modes and, therefore, provide data on the specific chemical bonds present yielding molecular 'IR fingerprints'. Being accurate, rapid, and inexpensive, these methods are particularly suitable for non-destructive quality assessments of a wide variety of agricultural products from grains and seeds to fruit and vegetables and also meat and dairy products.

Raman spectroscopy is another technique to determine vibrational modes of molecules. While NIR and MIR detect the absorption of specific bands of electromagnetic radiation, Raman spectroscopy relies on inelastic scattering of photons. The spectroscopic setup includes illuminating a sample with laser beam and collecting scattered electromagnetic radiation with a lens that is then filtered and dispersed onto a detector. With significantly lower cost of lasers nowadays, portable and inexpensive Raman spectrometers and Raman analyzers are becoming available, thus, broadening the application of this versatile spectroscopic method to food analysis. Similar to NIR and MIR, the method is widely applicable to a variety of agricultural commodities and food products. In some cases, Raman analysis could be carried out for food products inside the packages allowing for incorporation of such instruments into production lines or on-the-spot analysis of food products on a supermarket shelf. One of the main advantages of Raman spectroscopy over IR methods is that the results do not suffer from interference from water molecules, which typically present in food.

Nowadays other spectroscopic tools of modern analytical chemistry—namely UV–vis, NMR, and EPR spectroscopy—are also broadly applied to food analysis.

As one of the most powerful and informative structural techniques in chemistry, NMR, is based on detecting interactions of the radiofrequency field with nuclear spins, and is uniquely suited for reporting on molecular composition of food and food quality. While modern high resolution NMR instruments are rather expensive because of the necessity of employing superconductive magnets and specialized cryoprobes for the state-of-the-art resolution and sensitivity, inexpensive low field table-top NMR analyzers are now also available from several companies, and the number of such instruments is also growing. The latter instruments could be used for analyzing a broad range of food products. This volume contains a chapter that highlights NMR applications in the analysis of food emulsions, beverages, milk and dairy products, meats, carbohydrates, as well as fruit and vegetables.

Spectroscopic methods are also fully suitable for analyzing food products and agricultural commodities for trace compounds and contaminants such as herbicides and pesticides and/or carcinogen chemicals. This is becoming a very important task, as the public awareness of food contamination is growing and the regulatory agencies around the world are intensifying work on ensuring food safety. In addition to the topics mentioned earlier, this volume also provides an example of the use of a number of spectroscopic approaches to analyze the amount of ammonia nitrogen in aquaculture water. Ammonia nitrogen is an indispensable source of nitrogen for protein synthesis in aquatic organisms. However, at high concentrations ammonia is also toxic to aquatic organisms and would cause an undesirable eutrophication threatening aquaculture production. The chapter discusses the use of spectrophotometry, atomic absorption spectroscopy, infrared spectroscopy, fluorescence method and chromatography to assess ammonia nitrogen as well as challenges still present in this field.

Clearly, the individual contributions to this volume provide just a snapshot of the many applications of spectroscopic methods to food analysis. However, by no means should the presented list of topics be considered as an exhaustive review. The main intention of the volume was not only to merely highlight the individual spectroscopic methods by reviewing the current state of the art, but also to point out to the reader the existing problems and, in some cases, to indicate potential ways to resolve them. Thus, while I am very happy to present this volume, I am already looking forward to the new developments in the field of spectroscopy as applied to food science and food products.

Alex I Smirnov
Department of Chemistry, North Carolina State University,
2620 Yarbrough Drive, Raleigh, NC 27606-8205, USA
Alex_Smirnov@ncsu.edu

Editor biography

Ashutosh Kumar Shukla

With almost 19 years of physics teaching and research experience, publications in peer reviewed journals, review articles, textbooks and about 10 edited volumes published by reputed publishers, prepared in collaboration with experts from different countries, I enjoy academic work and intend to continue.

List of contributors

J Chapman
School of Science, RMIT University, Melbourne, Victoria, Australia

V K Truong
School of Science, RMIT University, Melbourne, Victoria, Australia

S Gangadoo
School of Science, RMIT University, Melbourne, Victoria, Australia

D Cozzolino
School of Science, RMIT University, Melbourne, Victoria, Australia

Bambang Kuswandi
Chemo and Biosensors Group, Faculty of Pharmacy, University of Jember, Indonesia

Y Parlak
Sakarya Applied Sciences University, Pamukova Vocational School, Turkey

Daoliang Li
College of Information and Electrical Engineering, China Agricultural University, China

Zhen Li
College of Information and Electrical Engineering, China Agricultural University, China

Cong Wang
College of Information and Electrical Engineering, China Agricultural University, China

Tan Wang
College of Information and Electrical Engineering, China Agricultural University, China

Xianbao Xu
College of Information and Electrical Engineering, China Agricultural University, China

Chapter 1

State of the art and applications of near infrared spectroscopy in food composition and quality

J Chapman, V K Truong, S Gangadoo and D Cozzolino

The food industry has embraced simple, rapid and cost-effective techniques for objectively evaluating and measuring the composition of raw materials, ingredients and foods to assure their quality to the consumer. Spectroscopic methods and techniques based in the visible (VIS), UV, near (NIR) and mid infrared (MIR) range of the electromagnetic spectrum have become an analytical option for the routine analysis of foods. Advantages of using UV–VIS, NIR, MIR and Raman spectroscopy combined with multivariate data analysis to analyse the composition and quality properties of raw materials and foods have been demonstrated by several authors. This chapter aims to highlight and present with recent applications of NIR spectroscopy for the analysis of foods.

1.1 Introduction

The definition of food quality generally refers to the degree to which a set of inherent characteristics fulfils requirements. Several physical and mechanical properties (e.g. mass, volume, sphericity, density and firmness), sensory attributes and properties (texture, taste and aroma), as well as their relationship with chemical composition are used to characterise and measure the so called 'food quality'. Other attributes of the food and raw ingredients, such as appearance, lack of defects, diseases (e.g. insects, fungus) are also used as proxies to quality (Nicolai *et al* 2007, Ayvaz *et al* 2015, 2016a, 2016b, Su and Sun 2018, Roberts *et al* 2018).

Advantages of using UV–VIS, NIR, MIR and Raman spectroscopy combined with multivariate data analysis to analyse the composition and quality properties of raw materials and foods have been demonstrated by several authors. The chapter is organized to provide a general introduction about the different vibrational spectroscopy methods commonly used in the analysis of food and food ingredients (e.g. VIS, NIR, MIR and Raman). A brief description of the multivariate data analysis

doi:10.1088/2053-2563/ab4428ch1

(MVA) methods used to analyse the data generated by these methods will be also presented. In addition, recent applications of NIR spectroscopy for the analysis of foods will be highlighted.

1.2 Spectroscopic techniques

1.2.1 Near and mid infrared spectroscopy

Methods based on vibrational spectroscopy [near (NIR) and mid (MIR)] have been widely used to measure the chemical composition of food ingredients and foods. These methods are based on the measurement of chemical bonds present in the sample that vibrate at specific frequencies, which are directly associated with the mass of the constituent atoms and the shape of the molecule (Woodcock *et al* 2008, Karoui *et al* 2010, Rodriguez-Saona and Allendorf 2011). Specific vibrational bonds absorbed in the infrared (IR) spectral region are related with diatomic molecules that only have one bond that may stretch (e.g. the distance between two atoms may increase or decrease) (McClure 2003, Blanco and Villaroya 2002).

The IR radiation belongs to the electromagnetic spectrum between the visible (VIS) and the microwave wavelengths (McClure 2003, Blanco and Villaroya 2002, Cozzolino 2009). The nominal range of wavelengths for NIR is between 750 and 2500 nm (13 400 to 4000 cm^{-1}), while for the MIR, it is from 2500 to 25 000 nm (4000 to 400 cm^{-1}) (McClure 2003, Blanco and Villaroya 2002, Cozzolino *et al* 2006, Cozzolino 2009).

Vibrational spectroscopy in the NIR range is characterised by low molar absorptions (McClure 2003, Blanco and Villaroya 2002, Cozzolino *et al* 2006, Cozzolino 2009). More broadly, NIR spectroscopy is capable of measuring organic chemical molecules containing O–H (oxygen–hydrogen), N–H (nitrogen–hydrogen) and C–H (carbon–hydrogen) bonds through the absorption of energy in the NIR region of the spectrum (Huang *et al* 2008, McClure 2003, Blanco and Villaroya 2002, Cozzolino *et al* 2006, Cozzolino 2009).

On the other hand, spectral 'signatures' in the MIR result from the fundamental stretching, bending, and rotating vibrations of the sample molecules, whilst NIR spectra result from complex overtones and high frequency combinations at shorter wavelengths (McClure 2003, Blanco and Villaroya 2002, Cozzolino *et al* 2006, Roggo *et al* 2007, Cozzolino 2009). Although NIR intensities are 10–1000 times lower than for those in the MIR region, highly sensitive spectrometers can be built through several means including the use of efficient detectors and brighter light sources (McClure 2003, Blanco and Villaroya 2002, Roggo *et al* 2007, Cozzolino 2009).

Peaks in the MIR range are often sharper and better resolved than in the NIR region, where higher overtones of vibrations derived from the O–H, N–H, C–H and S–H (sulphur–hydrogen) bands from the MIR range are still observed in the NIR region, although much weaker than the fundamental frequencies in the MIR. In addition to the existence of combination bands (e.g. CO stretch and NH bend in protein), it gives rise to NIR spectrum with strongly overlapping bands (McClure 2003, Nicolai *et al* 2007, Cozzolino *et al* 2006, Cozzolino 2009). A major disadvantage of using NIR spectroscopy is that both band overlapping and complexity of the spectra has been very difficult for a direct quantification and interpretation of the data generated during the analysis of samples using NIR spectroscopy. Yet, the broad overlapping bands in the NIR region can reduce the need for using several wavelengths during calibration development and routine analysis. In recent years, new instrumentation and computer algorithms have taken advantage of this complexity and have made the technique much more powerful and simple to use (Nicolai *et al* 2007, Cozzolino 2009). Yet, the advent of inexpensive and powerful computers has contributed to the surge of new NIR applications (Nicolai *et al* 2007, Cozzolino 2009). Figure 1.1 shows an example of the NIR and MIR spectra of a food indicating the main regions of interest used for the analysis of composition in foods.

1.2.2 Raman spectroscopy

Infrared spectroscopy (IR) has been used in different applications in the food industry, but its main disadvantage is associated with its intrinsic limitation, the strong effect of the absorption bands of water (Ozaki and Sasic 2008). It is well known that Raman spectroscopy appears more suitable for the vibrational assessment of water-containing samples, due to the relatively weak water bending mode (Ozaki and Sasic 2008). Raman spectroscopy, based on the inelastic light scattering on molecules, is gaining an increasingly wider area of applications. The inelastic scattered light by the molecule provides rich information on its vibrational modes, thus being characteristic to each molecular structure (Ozaki and Sasic 2008). Although a small amount of the incident light is inelastically scattered, where the so-called Raman effect is an inherently weak effect, the weak counterpart of the incident light energy is modulated by the molecular vibrations of the scattering sample (Ozaki and Sasic 2008). Therefore, the observed vibrational responses provide enough information about the chemical composition of the food sample. In recent years, measuring the Raman spectra of complex samples *in situ* became accessible once the high resolution portable instruments with optimised detection capability were released on the market (Ozaki and Sasic 2008).

1.2.3 UV and visible spectroscopy

UV–VIS spectroscopy is characterised by the physical responses (electronic transitions) of food molecules to light. The physical responses are originated by absorption, scattering, diffraction, refraction, and reflection (Sommer 2012, Perkampus 1992, Parmar and Sharma 2016). The phenomenon of UV and VIS light absorption is restricted to specific chromophores and several chemical species with defined molecular functional groups (Sommer 2012, Perkampus 1992, Parmar and Sharma 2016). Subsequently, the characteristic UV–VIS spectra might originate from single molecules as electrons within these chromophores are excited during the analysis (Sommer 2012, Perkampus 1992, Parmar and Sharma 2016). Quantitative analysis based on UV–VIS spectroscopy is ultimately described by the Beer–Lambert law, where the correlation between the quantity of the incident light

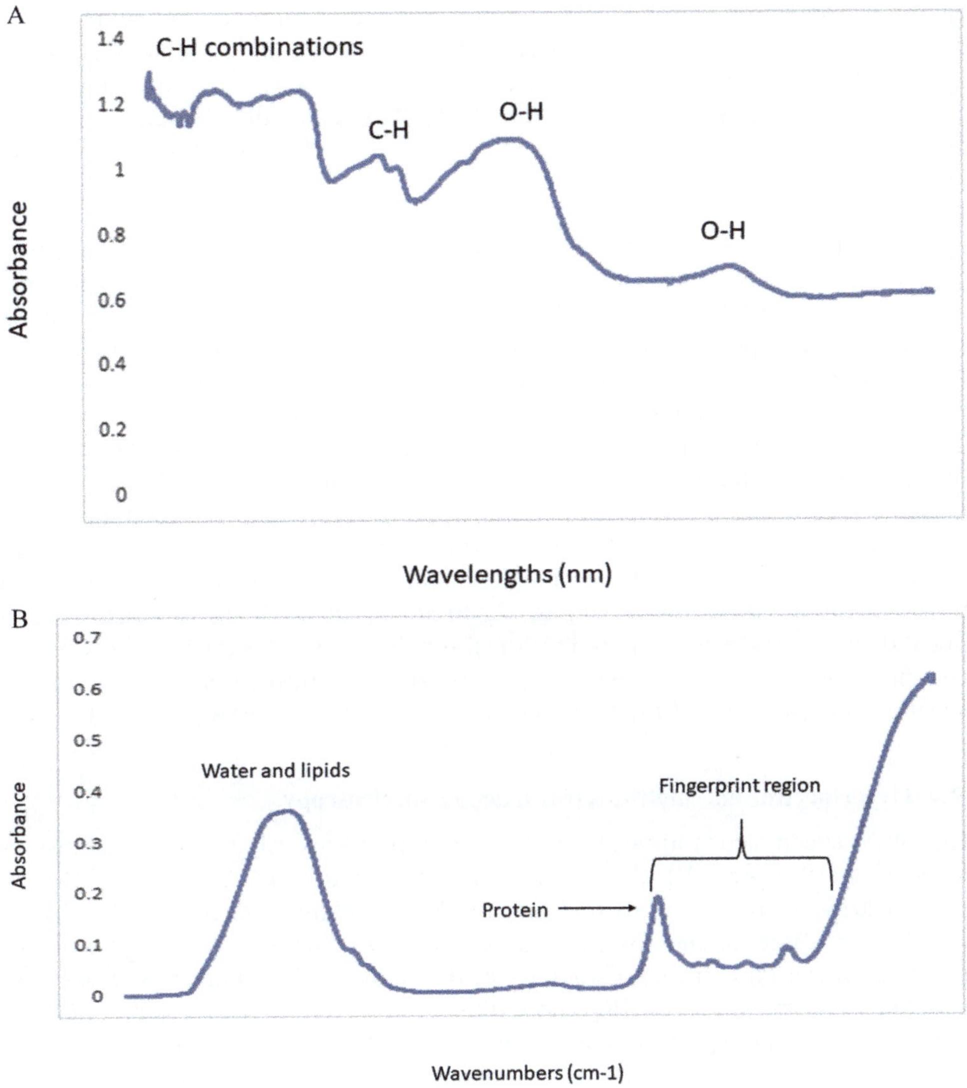

Figure 1.1. Example of the NIR (A) and MIR (B) spectra of a food indicating the main regions of interest used for the analysis of composition in foods.

absorbed by the molecule, the sample, the light path length and the concentration of the absorbing compound or molecule in the matrix is measured (Sommer 2012, Perkampus 1992, Parmar and Sharma 2016). Methods based on the use of UV–VIS allowed for the determination and quantification of the target molecule concentration within the food matrix (Sommer 2012, Perkampus 1992, Parmar and Sharma 2016). In recent years, the continuous development of optical instruments has improved the analytical capabilities of UV–VIS spectrophotometers. For example, the use of fibre optics in combination with linear photodiode arrays or charge-coupled devices as detectors provided more portable and compact instruments. In addition, spectrophotometers have become highly sensitive and capable of detecting low concentrations in complex matrices (e.g. aqueous solutions) (Sommer 2012, Perkampus 1992, Parmar and Sharma 2016). Commercially available UV–VIS spectrometers capable of quantitative analysis of chemical and biological samples are designed for liquid samples (Sommer 2012, Perkampus 1992, Parmar and Sharma 2016). Solid samples are not easily dissolved and often require a significant level of sample preparation, such as complete dissolution in an appropriate solvent, where more aggressive approaches, such as acid digestion, are performed (Sommer 2012, Perkampus 1992, Parmar and Sharma 2016). The analysis of UV absorption solutes can only be performed in homogenous solutions, as solid particles are present in non-homogeneous samples which contributes to significant interference observed within the spectra, due to the absorption and light scattering effects of the individual particles. In the biotechnology and food industries, the phenomenon of non-homogeneity is exploited in optical density (OD) measurements, which can then be used to determine the biomass concentration in turbid samples (Sommer 2012, Perkampus 1992, Parmar and Sharma 2016). It should be noted that for proteins and other similarly large molecules, incident light is absorbed by multiple functional groups within the compound, which results in non-specific UV–VIS spectra. Several optical density sensors have been developed for the food industry that allow for in-line applications, where the sensors are based on transmission or turbidimetry measurements (Sommer 2012, Perkampus 1992, Parmar and Sharma 2016).

1.2.4 Hyperspectral and multispectral imaging spectroscopy

Most of the spectroscopy applications have been primarily relying on spot measurements using either the NIR or MIR range (Roberts *et al* 2018, Clark *et al* 2019, Su and Sun 2018, Pallone *et al* 2018, Huang *et al* 2017, Wang *et al* 2016, 2018). In the last 20 years, cameras and spectral imaging devices, now readily available on the market provide exciting new possibilities for the analysis of food samples. The recent availability of hyperspectral (HSI) and multispectral imaging (MSI) systems allowed for obtaining spatial, spectral, and multi-constituent information about the sample being analysed (Karoui *et al* 2010, Rodriguez-Saona and Allendorf 2011, Roberts *et al* 2018, Clark *et al* 2019, Su and Sun 2018, Pallone *et al* 2018, Huang *et al* 2017, Wang *et al* 2016, 2018). In the last decade, both HSI and MSI systems have started to feature with increasing prominence as rapid and efficient methods for the assessment of food quality and integrity (e.g. origin, microbial contamination, food safety, composition) (Karoui *et al* 2010, Rodriguez-Saona and Allendorf 2011, Roberts *et al* 2018, Clark *et al* 2019, Su and Sun 2018, Pallone *et al* 2018). Spectral imaging can be classified as either HSI or MSI (Karoui *et al* 2010, Rodriguez-Saona and Allendorf 2011, Roberts *et al* 2018, Clark *et al* 2019, Su and Sun 2018, Pallone *et al* 2018). Multispectral imaging involves the acquisition of spectral images at few discrete and narrow wavebands (bandwidths of between 5 and 50 nm) and it is considered to be an improvement of hyperspectral imaging as this technology is cost effective (Karoui *et al* 2010, Rodriguez-Saona and Allendorf 2011, Roberts *et al* 2018,

Clark *et al* 2019, Su and Sun 2018, Pallone *et al* 2018). The capability of MSI to simultaneously predict multiple components, provides this technique with a key advantage and renders a promising outlook with a single, automated image acquisition (Karoui *et al* 2010, Rodriguez-Saona and Allendorf 2011, Roberts *et al* 2018, Clark *et al* 2019, Su and Sun 2018, Pallone *et al* 2018). Hyperspectral imaging relies on one of two sensing modes as in-line scanning (push broom) mode or as filter-based imaging mode (Karoui *et al* 2010, Rodriguez-Saona and Allendorf 2011, Roberts *et al* 2018, Clark *et al* 2019, Su and Sun 2018, Pallone *et al* 2018, Huang *et al* 2017, Wang *et al* 2016, Wang *et al* 2018). Methods based in HIS require a high performance digital camera that can cover the spectral region of interest, in a wide dynamic range with high signal-to-noise-ratio, and good quantum efficiency (Karoui *et al* 2010, Rodriguez-Saona and Allendorf 2011, Roberts *et al* 2018, Clark *et al* 2019, Su and Sun 2018, Pallone *et al* 2018, Huang *et al* 2017, Wang *et al* 2016, 2018). Hyperspectral imaging is particularly being used in food because of the integration of imaging ability with spectroscopy (e.g. NIR, MIR and Raman), which enables the simultaneous acquisition of both spectral and spatial information from the target (Roberts *et al* 2018, Clark *et al* 2019, Su and Sun 2018, Pallone *et al* 2018, Elmasry *et al* 2012, Huang *et al* 2017, Wang *et al* 2016, 2018).

1.3 Multivariate data analysis

Most of the modern techniques described in the sections above generate vast amounts of data where the application of these techniques in the food industry goes hand in hand with the combination of multivariate data methods (Williams *et al* 2017, Roberts and Cozzolino 2016). Both quantitative and qualitative tools are used to extract information form the data generated by these instruments. In general terms, either regression or calibration are defined as the most common methods used to fit a model to a set of observed data, to quantify the relationships between variables (quantitative analysis) (Brereton 2008, Esbensen 2002, Williams *et al* 2017). There are several steps to be considered in the calibration process (e.g. sample selection, reference analysis, calibration and validation). Overall, the selection of samples (sampling) is perhaps the most underestimated part of the calibration process, having a major effect in the interpretation of the calibration (Cozzolino 2015a, 2015b, Murray 1999, Murray and Cowe 2004). Any robust model or calibration involves the careful selection of the samples into the calibration set and the selection of samples to validate or test the calibration model (Cozzolino 2015a, 2015b, Murray 1999, Murray and Cowe 2004, Williams *et al* 2017). Samples selected for calibration must span the whole range of variability in both the calibration and validation sets and have the same treatments or pre-processing to be followed with future samples (Cozzolino 2015a, 2015b, Murray 1999, Murray and Cowe 2004).

Calibration is a process involving the use of regression techniques coupled with pre-processing methods (Brereton 2008, Naes *et al* 2002, Nicolai *et al* 2007, Williams *et al* 2017). During the development of an application based on the use of vibrational spectroscopy, the aim of regression analysis is to build a model that relates the information of the set of known measurements (e.g. spectral features) to the desired property (usually an analytic measurement from a reputable reference method) (Naes *et al* 2002, Nicolai *et al* 2007, Walsh and Kawano 2009, Williams *et al* 2017).

A wide range of algorithms become available to develop regression models where the most widely used are least squares regression (LSR), multiple linear regression (MLR), partial least squares (PLS) and principal component regression (PCR) (Naes *et al* 2002, Williams *et al* 2017). Before regression, the spectral data often go through different pre-processing steps (e.g. scattering or baseline corrections, first or second derivative calculations with spectral smoothing, or straight-line subtraction) allowing the signal from dissimilar samples to be compared. The choice of the appropriate method to be applied is highly dependent on the sample, instrument, and purpose of analysis. Several pre-processing methods are available for this purpose where specific details can be found elsewhere (Blanco and Bernardez 2009, Naes *et al* 2002, Nicolai *et al* 2007).

After the model is developed, its ability to predict unknown samples withheld from the calibration set (e.g. not used to construct the model) should be tested (Nicolai *et al* 2007, Walsh and Kawano 2009, Westad and Marini 2015, Williams *et al* 2017). This step requires the use of the model to predict the targeted property in a restricted number of samples not included in the calibration or training (Westad and Marini 2015, Williams *et al* 2017). The results obtained after the validation of the model should be directly compared with the actual reference values; if the two agree (in the best-case scenario they are identical), the model will meet the requirements to accurately predict the property in the future (Naes *et al* 2002, Nicolai *et al* 2007, Walsh and Kawano 2009, Williams *et al* 2017). Overall, the predictive ability of a model must be demonstrated using an independent set of samples where an independent sample should be sourced from other experiments, batches or conditions (Norris and Ritchie 2008, Naes *et al* 2002, Nicolai *et al* 2007, Walsh and Kawano 2009, Cozzolino 2015a, 2015b, Murray and Cowe 2004, Williams *et al* 2017).

Several statistics have been described to interpret and report the results obtained during model development and validation (Norris and Ritchie 2008, Williams *et al* 2017). Some of them include the prediction error of a calibration model defined as the root mean square error for cross validation (RMSECV), the root mean square error for prediction (RMSEP), the residual predictive deviation (RPD) value (defined as the ratio of the standard deviation of the response variable to the RMSEP or RMSECV (some authors use the term SDR)), and coefficient of determination (R2) (Fearn 2002, Williams 2001, Norris and Ritchie 2008, Williams *et al* 2017).

After the model is developed, the fit-for-purpose criterion needs to be considered to judge its suitability for routine use (Williams *et al* 2017). Food scientists and technologists need to interpret the models in the overall context of the application and not only on the cold interpretation of the statistics (Fearn 2002, Williams 2001, Williams *et al* 2017).

1.4 Applications

Table 1.1 summarises current applications of NIR spectroscopy for the analysis of composition in different foods (e.g. cereals, meat, fruit and vegetables, beverages). The reader is guided to read the large amount of information available in the field. Overall, NIR spectroscopy can be used to analyse the main composition of foods such as protein, carbohydrates, lipids/fat and water (dry matter). The last 20 years have expanded the applications of NIR spectroscopy to other chemical and physical parameters used to characterise foods (figure 1.2). The measurement of antioxidants, colour, phenolic and volatile compounds in fruit and beverages, sensory properties and mechanical characteristics (tenderness in meat), non-destructive analysis of cheese and dairy products, olive oil quality, milk composition, are a few examples (Cozzolino *et al* 2006, Gishen *et al* 2010, Nicolai *et al* 2007, Chapman *et al* 2019, Texeira do Santos *et al* 2017a, Texeira do Santos *et al* 2017b, Prieto *et al* 2009, Holroyd 2013, Gomez-Caravaca *et al* 2016, Zhang *et al* 2017, Sileoni *et al* 2017, Lei and Sun 2019, Caporaso *et al* 2018, Cortés *et al* 2019, Hashimoto and Kameoka 2008, McGoverin et al 2010). Near infrared spectroscopy has been also explored as rapid phenotyping method during breeding of cereals, fruits and vegetables (Pojić and Mastilović 2012, Ayvaz *et al* 2015, 2016a, 2016b, Cozzolino and Roberts 2016).

Table 1.1. Examples of current applications of NIR spectroscopy methods in the analysis of food ingredients and foods.

Parameter or analyte	Matrix, food ingredient or food analysed
Protein or nitrogen	Cereals
	Milk
	Meat
	Food ingredients
Fat or lipids	Cereals
	Milk
	Meat
Moisture or dry matter	Cereals
	Milk
	Meat
	Food ingredients
Alcohol content	Wine
	Beer
	Other alcoholic beverages
Sugars	Cereals
	Wine
	Beer
pH and titratable acidity	Wine
	Beer
	Fruits

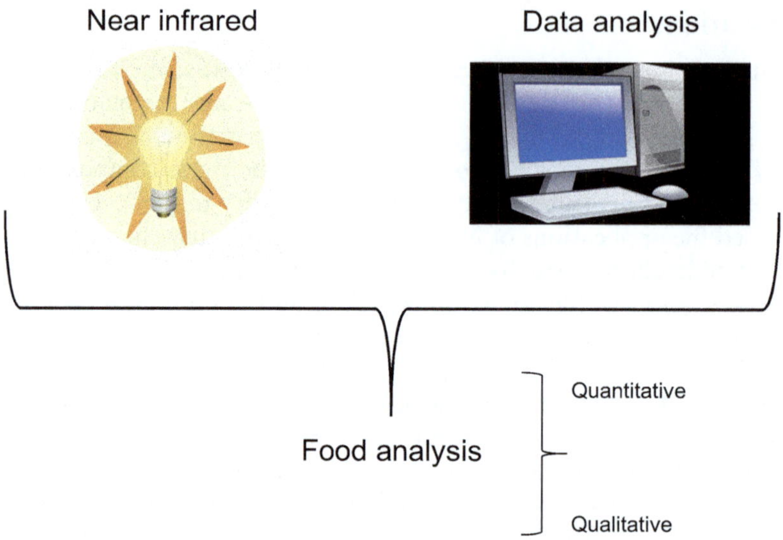

Figure 1.2. Combining near infrared spectroscopy and data analysis for the routine analysis of foods. The lightbulb image has been obtained by the authors from humbliceous.blogspot.com where it was made available under a CC BY-SA 2.5 licence. It is included within this chapter on that basis. It is attributed to Curioso.

More recently, NIR spectroscopy has been used as a method to target issues related with authentication, fraud, origin and traceability of foods (Cozzolino 2014, 2015a, 2015b, Ellis *et al* 2012, Teixeira dos Santos *et al* 2017a, 2017b, Urícková and Sádecká 2015, Lohumi *et al* 2015, Callao and Ruisanchez 2018) and to monitor food processing (Grassi and Alamprese 2018).

1.5 Summary

This chapter highlighted and briefly presented some of the most common spectroscopy methods used in food analysis, with special emphasis in NIR spectroscopy applications. In addition, a brief description of the use of multivariate analysis as a tool to enhance the application of such methods was provided. New developments in instruments (e.g. miniaturisation and hand-held spectrophotometers) have determined a reduction in the cost of the instruments and are no longer bound to the laboratory. Unfortunately, few studies have evaluated the performance of portable systems under 'field' situations.

The characteristics and properties of the sample can present challenges for the implementation of these methods, as the sample can be unsuitable for real world quantitative spectroscopic analysis (e.g. pre-processing will be required). The analytical measurement might be hindered by different factors such as the complex nature of the sample, inaccurate measurements occurring due to interferences or masking agents, and the inability of the analyst to interpret directly the results generated during the analysis.

More recently, vibrational spectroscopy has been evaluated in automation or process control (in line, on line) due to the advantages of these methods including a diverse range in sample presentation alternatives, the incorporation of fibre optics, portable instruments, among other characteristics. However, bench top and laboratory instruments are well-embedded throughout the food industry as methods for quality control.

References

Ayvaz H, Bozdogan A, Giusti M M, Mortas M, Gomez R and Rodriguez-Saona L E 2016a Improving the screening of potato breeding lines for specific nutritional traits using portable mid-infrared spectroscopy and multivariate analysis *Food Chem.* **211** 374–82

Ayvaz H, Santos A M, Moyseenko J, Kleinhenz M and Rodriguez-Saona L E 2015 Application of a portable infrared instrument for simultaneous analysis of sugars, asparagine and glutamine levels in raw potato tubers *Plant Foods Hum. Nutr.* **70** 215–20

Ayvaz H, Sierra-Cadavid A, Aykas D P, Mulqueeney B, Sullivan S and Rodriguez-Saona L E 2016b Monitoring multicomponent quality traits in tomato juice using portable mid-infrared (MIR) spectroscopy and multivariate analysis *Food Control* **66** 79–86

Blanco M and Bernardez M 2009 Multivariate calibration for quantitative analysis *Infrared Spectroscopy for Food Quality Analysis and Control* ed D W Sun (Oxford: Elsevier)

Blanco M and Villaroya I 2002 NIR spectroscopy: a rapid-response analytical tool *Trends Anal. Chem.* **21** 240

Brereton R G 2008 *Applied Chemometrics for Scientists* (Chichester: Wiley)

Callao M P and Ruisanchez I 2018 An overview of multivariate qualitative methods for food fraud detection *Food Control* **6** 283–93

Caporaso N, Whitworth M B and Fisk I D 2018 Near-Infrared spectroscopy and hyperspectral imaging for non-destructive quality assessment of cereal grains *Appl. Spectrosc. Rev.* **53** 667–87

Chapman J, Elbourne A, Truong V K, Newman L, Gangadoo S, Rajapaksha Pathirannahalage P, Cheeseman S and Cozzolino D 2019 Sensomics—from conventional to functional NIR spectroscopy—shining light over the aroma and taste of foods *Trends Food Sci. Technol.* **91** 274–81

Clark C, Cozzolino D and Bureau S 2019 Contributions of Fourier-transform mid infrared (FT-MIR) spectroscopy to the study of fruit and vegetables: A review *Postharvest Biol. Technol.* **148** 1–14

Cortés V, Blasco J, Aleixos N, Cubero S and Talensa P 2019 Monitoring strategies for quality control of agricultural products using visible and near-infrared spectroscopy: A review *Trends Food Sci. Technol.* **85** 138–48

Cozzolino D 2009 Near infrared spectroscopy in natural products analysis *Planta Med.* **75** 746–57

Cozzolino D 2014 An overview of the use of infrared spectroscopy and chemometrics in authenticity and traceability of cereals *Food Res. Int.* **60** 262–5

Cozzolino D 2015a The role of vibrational spectroscopy as tool to assess economical motivated fraud and counterfeit issues in agricultural products and foods *Anal. Methods* **7** 9390–400

Cozzolino D 2015b The role of visible and infrared spectroscopy combined with chemometrics to measure phenolic compounds in grape and wine samples *Molecules* **20** 726–35

Cozzolino D *et al* 2006 Chemometrics and visible-near infrared spectroscopic monitoring of red wine fermentation in a pilot scale *Biotechnol. Bioeng.* **95** 1101–7

Cozzolino D and Roberts J J 2016 Applications and developments on the use of vibrational spectroscopy imaging for the analysis, monitoring and characterisation of crops and plants *Molecules* **21** 755–63

Ellis D I, Brewster D L, Dunn W B, Alwood J W, Golovanov A P and Goodacre R 2012 Fingerprinting food: current technologies for the detection of food adulteration and contamination *Chem. Soc. Rev.* **41** 5706–27

Elmasry G, Barbin D F, Sun D W and Allen P 2012 Meat quality evaluation by hyperspectral imaging technique: an overview *Crit. Rev. Food Sci. Nutr.* **52** 689–711

Esbensen K H 2002 *Multivariate Data Analysis in Practice* (Oslo: CAMO Process AS)

Fearn T 2002 Assessing calibrations: SEP, RPD, RER and R2 *NIR News* **13** 12–4

Gishen M, Cozzolino D and Dambergs R G 2010 *Applications of Vibrational Spectroscopy in Food Science* ed E Li-Chan, J Chalmers and P Griffiths (New York: Wiley)

Gomez-Caravaca A M, Maggio R M and Cerretani L 2016 Chemometric applications to assess quality and critical parameters of virgin and extra-virgin olive oil. A review *Anal. Chim. Acta* **913** 1–21

Grassi S and Alamprese C 2018 Advances in NIR spectroscopy applied to process analytical technology in food industries *Curr. Opin. Food Sci.* **22** 17–21

Hashimoto A and Kameoka T 2008 Applications of infrared spectroscopy to biochemical, food, and agricultural processes *Appl. Spectrosc. Rev.* **43** 416–51

Holroyd S E 2013 The use of near infrared spectroscopy on milk and milk products *J. Near Infrared Spectrosc.* **21** 311–22

Huang H, Yu H, Xu H and Ying Y 2008 Near infrared spectroscopy for on/in-line monitoring of quality in foods and beverages: a review *J. Food Eng.* **87** 303–13

Huang Y, Lu R and Chen K 2017 Development of a multichannel hyperspectral imaging probe for property and quality assessment of horticultural products *Postharvest Biol. Technol.* **133** 88–97

Karoui R, Downey G and Blecker C 2010 Mid-infrared spectroscopy coupled with chemometrics: a tool for the analysis of intact food systems and the exploration of their molecular structure–quality relationships – a review *Chem. Rev.* **110** 6144–68

Lei T and Sun D-W 2019 Developments of nondestructive techniques for evaluating quality attributes of cheeses: A review *Trends Food Sci. Technol.* **88** 527–42

Lohumi S, Lee S, Lee H and Cho B-W 2015 A review of vibrational spectroscopic techniques for the detection of food authenticity and adulteration *Trends Food Sci. Technol.* **46** 85–98

McClure F W 2003 204 years of near infrared technology: 1800 – 2003 *J. Near Infrared Spectrosc.* **11** 487–8

McGoverin C M, Weeranantanaphan J, Downey G and Manley M 2010 The application of near infrared spectroscopy to the measurement of bioactive compounds in food commodities *J. Near Infrared Spectrosc.* **18** 87–111

Murray I 1999 NIR spectroscopy of food: simple things, subtle things and spectra *NIR News* **10** 10–2

Murray I and Cowe I 2004 Sample preparation *Near Infrared Spectroscopy in Agriculture* ed C A Roberts, J Workman and J B Reeves (Madison, WI: American Society of Agronomy, Crop Science Society of America, Soil Science Society of America), pp 75–115

Naes T, Isaksson T, Fearn T and Davies T 2002 *A User-friendly Guide to Multivariate Calibration and Classification* (Chichester: NIR Publications), 420 pp

Nicolai B M, Beullens K, Bobelyn E, Peirs A, Saeys W, Theron K I and Lammertyn J 2007 Nondestructive measurement of fruit and vegetable quality by means of NIR spectroscopy: A review *Postharvest Biol. Technol.* **46** 99–118

Norris K H and Ritchie G E 2008 Assuring specificity for a multivariate near-infrared (NIR) calibration: The example of the Chambersburg Shoot-out 2002 data set *J. Pharm. Biomed. Anal.* **48** 1037–41

Ozaki Y and Sasic S 2008 Introduction to Raman spectroscopy *Pharmaceutical Applications of Raman Spectroscopy* ed S Sasic (Hoboken, NJ: Wiley), pp 1–28

Pallone J A, dos Santos Carames E M and Domingues Alamar P 2018 Green analytical chemistry applied in food analysis: alterative techniques *Curr. Opin. Food Sci.* **22** 115–21

Parmar A and Sharma S 2016 Derivative UV-vis absorption spectra as an invigorated spectrophotometric method for spectral resolution and quantitative analysis: Theoretical aspects and analytical applications: A review *Trends Anal. Chem.* **77** 44–53

Perkampus H-H 1992 Recent developments in UV-VIS spectroscopy *UV-VIS Spectroscopy and Its Applications* (Berlin: Springer), pp 81–130

Pojić M M and Mastilović J S 2012 Near infrared spectroscopy—advanced analytical tool in wheat breeding, trade, and processing *Food Bioprocess Technol.* **6** 330–52

Prieto N, Roehe R, Lavín P, Batten G and Andrés S 2009 Application of near infrared reflectance spectroscopy to predict meat and meat products quality: A review *Meat Sci.* **83** 175–86

Roberts J J and Cozzolino D 2016 An overview on the application of chemometrics in food science and technology—an approach to quantitative data analysis *Food Anal. Methods* **9** 3258–67

Roberts J, Power A, Chapman J, Chandra S and Cozzolino D 2018 Vibrational spectroscopy methods for agro-food product analysis *Compr. Anal. Chem.* **80** 51–68

Roberts J, Power A, Chapman J, Chandra S and Cozzolino D 2018 A short update on the advantages, applications and limitations of hyperspectral and chemical imaging in food authentication *Appl. Sci.* **8** 505–10

Roberts J, Power A, Chapman J, Chandra S and Cozzolino D 2018 The use of UV-Vis spectroscopy in bioprocess and fermentation monitoring *Fermentation* **4** 18

Rodriguez-Saona L E and Allendorf M E 2011 Use of FTIR for rapid authentication and detection of adulteration of food *Annu. Rev. Food Sci. Technol.* **2** 467–83

Roggo Y, Chalus P, Maurer L, Lema-Martinez C, Edmond A and Jent N 2007 A review of near infrared spectroscopy and chemometrics in pharmaceutical technologies *J. Pharm. Biomed. Anal.* **44** 683–700

Sileoni V, Marconi O and Perretti G 2017 Near-infrared spectroscopy in the brewing industry *Crit. Rev. Food Sci. Nutr.* **55** 1771–91

Sommer L 2012 *Analytical Absorption Spectrophotometry in the Visible and Ultraviolet: The Principles* vol 8 (Amsterdam: Elsevier)

Su W-H and Sun D-W 2018 Multispectral imaging for plant food quality analysis and visualization *Compr. Rev. Food Sci. Food Safety* **17** 220–39

Teixeira dos Santos C A, Pasco R N M J and Lopes J A 2017a A review on the application of vibrational spectroscopy in the wine industry: From soil to bottle *Trends Anal. Chem.* **88** 100–18

Teixeira dos Santos C A, Pásco R N M J, Cruz Sarraguça M, Porto P A L S, Cerdeira A L, González-Sáiz J M, Pizarro C and Lopes J A 2017b Merging vibrational spectroscopic data for wine classification according to the geographic origin *Food Res. Int.* **102** 504–10

Urícková V and Sádecká J 2015 Determination of geographical origin of alcoholic beverages using ultraviolet, visible and infrared spectroscopy: A review *Spectrochim. Acta Part A: Mol. Biomol. Spectr.* **148** 131–7

Walsh K B and Kawano S 2009 Near infrared spectroscopy *Optical Monitoring of Fresh and Processed Agricultural Crops* ed M Zude (Boca Raton, FL: CRC Press), pp 192–239

Wang N-N *et al* 2016 Recent advances in the application of hyperspectral imaging for evaluating fruit quality *Food Anal. Methods* **9** 178–91

Wang W, Peng Y and Sun H 2018 Spectral detection techniques for non-destructively monitoring the quality, safety, and classification of fresh red meat *Food Anal. Methods* **11** 2707–30

Westad F and Marini F 2015 Validation of chemometric models: A tutorial *Anal. Chim. Acta* **893** 14–23

Williams P C 2001 Implementation of near-infrared technology *Near Infrared Technology in the Agriculture and Food Industries* 2nd ed P Williams and K Norris (St Paul, MN: American Association of Cereal Chemists), pp 145–69

Williams P, Dardenne P and Flinn P 2017 Tutorial: Items to be included in a report on a near infrared spectroscopy project *J. Near Infrared Spectrosc.* **25** 85–90

Woodcock T, Downey G and O'Donnell C P 2008 Better quality food and beverages: the role of near infrared spectroscopy *J. Near Infrared Spectrosc.* **16** 1–29

Zhang B, Dai D, Huang J, Zhou J, Gui Q and Dai F 2017 Influence of physical and biological variability and solution methods in fruit and vegetable quality nondestructive inspection by using imaging and near-infrared spectroscopy techniques: A review *Crit. Rev. Food Sci. Nutr.* **58** 2099–118

Chapter 2

Food authenticity and traceability: molecular spectroscopy approaches

Bambang Kuswandi

Currently, food quality is becoming a major concern worldwide, due to food adulteration that could threaten consumer health and cause financial loss. In this regard, food characterization techniques have been helpful tools, especially molecular spectroscopy, because they are rapid, non-destructive, have simple operation and less or no need for sample preparation, so do not create toxic residues. These spectroscopy tools are UV–vis, infrared, Raman, and NMR (nuclear magnetic resonance) spectroscopy. Since the molecular spectra are mostly very complex and overlapping, this makes them less sensitive and selective. Therefore, chemometrics can be used to tackle this problem. This chapter reviews the current molecular spectroscopy development in conjunction with chemometrics for food characterization and quality analysis.

2.1 Introduction

Currently, consumer concern about the quality of food is increasing, perhaps due to consumer behavior changes such as eating habits, the supply chain, and the effect of the rapid increase of food industrialization. The need for food products with high quality and safety means a requirement for high standards of quality and process control with suitable and reliable analytical tools for research and testing. Food quality analysis is the analytical process that determines food quality. Food quality can be described as the minimum standards of components that qualify as suitable to be consumed by a human or allowed with food contact. Physical food quality, such as color, appearance, flavor and texture, are also crucial factors for food sensory quality. The food quality can be categorized as chemical, biological and microbial. For example, the degradation of food products that cause short shelf life is associated with irreversible enzymatic and chemical reactions (Otles and Ozyurt 2015). This is due to the fact that food is a complex matrix consisting mainly of

doi:10.1088/2053-2563/ab4428ch2

water, carbohydrates, proteins, and fat along with various minor elements. Therefore, their molecular structure and interactions within the food matrix system govern the functional properties of these food components, and their amounts present determine the food quality products characteristics.

Food quality has been the main issue of producers, consumers, and regulators since ancient times (Kuswandi *et al* 2011, Dobrucka and Cierpiszewski 2014). Sophisticated instrumentation techniques, advances in analytical methods and ICT (information and communication technologies) created analytical tools for precise food quality measurements (Biji *et al* 2015). Food quality analysis is a critical issue in food quality control and safety. Food quality analysis is a quality food criterion and ingredients based on food legislative protection. Therefore, there is a critical requirement for food quality techniques for accurate standardization (Osborne 2000, Abbas *et al* 2013). Food quality analysis is not only needed by consumers but also other stakeholders, such as food producers and industries who are searching for the chance to give assurance about their food products and branding. At the same time, the authorities are needed for an updated and extended the list of the analytical tools for food quality products confirmation to maintain law enforcement (Kuswandi 2016).

Molecular spectroscopic methods have been very reliable tools for the success of food product quality analysis. They are highly suitable for food components analysis since they often need less or even no sample preparation, provide fast and on-line analysis as well as affording performance of multiple analytes on a single sample. These benefits especially relate to NIR (near-infrared) and IR (infrared) and NMR (nuclear magnetic resonance) spectroscopy. The first technique is mainly employed as a quality assurance tool to determine food ingredients, intermediate processes, and finished products (Abbas *et al* 2013). In addition, UV–vis, fluorescence, and mid-infrared (MIR), as well as Raman spectroscopy, are applied in the monitoring of food quality.

The objective of this chapter is to describe applicability of molecular spectroscopic techniques, such as UV–vis, fluorescence, infrared, Raman and NMR spectroscopy, as rapid tools for food products quality analysis. In addition, the basic principles of the aforementioned techniques are presented combined with chemometric methods along with the research trends using spectroscopic tools in this food area.

2.2 Molecular spectroscopy

In molecular spectroscopy, the spectra obtained are originated from the absorption, emission, and scattering of a proton when the molecule energy changes (Skoog *et al* 1998). In molecules, there are energy states corresponding to nuclei vibrations and rotations. Generally, transitions within the rotation energy levels belong to the far infrared and microwave spectral region, energy transitions within the vibrational energy levels belong to the infrared spectral region, and the electronic energy transitions levels belong to the visible or ultraviolet spectral region, as given in figure 2.1, representing the regions of the electromagnetic spectrum. Therefore, the

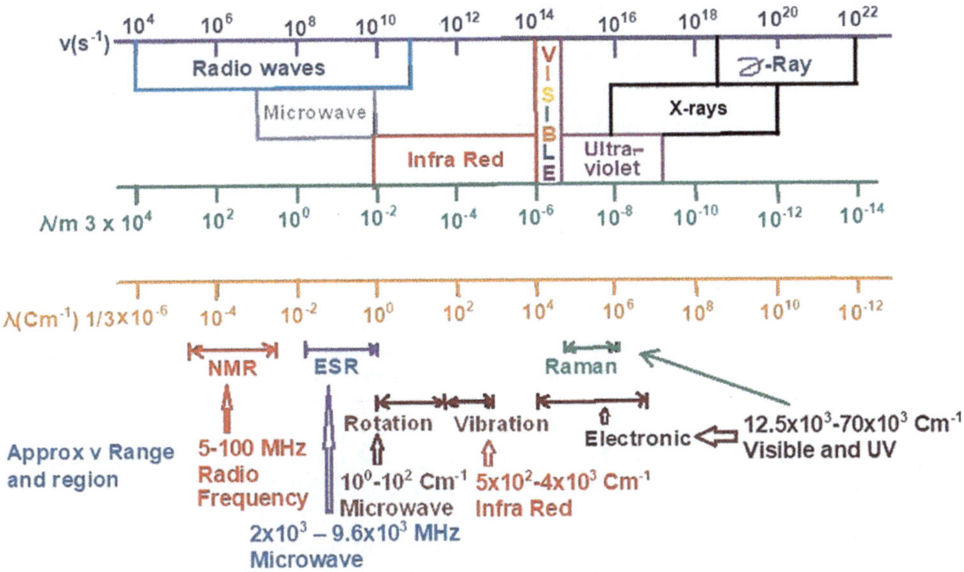

Figure 2.1. Regions of the electromagnetic spectrum.

molecular spectra are complex, since they contain information on the bond strength and the molecular structure (Oppermann 2003). Furthermore, they provide a variety of molecular properties that can be determined in many ways. Hence, the molecular spectroscopy is important for the investigation of properties of microscopic atomic and molecular objects, such as in food analysis.

The broadest range of application is captured by the UV–vis spectrometers that are an excellent tool for the analysis of food samples qualitatively and quantitatively in absorbance, transmittance and reflectance modes. Sensitive measurements of trace and ultra-trace elements in food samples solutions can be performed using these spectroscopic techniques. In addition, fluorescence spectroscopy is enabled for the quantitative analysis of food products with high sensitivity, completing the range of spectroscopic tools reliable for the food sector. Moreover, FTIR spectrometers, from the NIR to the FIR region, afford specific identification of food substances (Stuart 2004, Abbas *et al* 2013). These techniques could be applied along with a wide range of image technologies, including microscopy; each technology is selected based on the food substances and characteristics.

Hence, the molecular spectrometers are the appropriate tools for measuring a varying range of food analysis applications, including food contamination and adulteration. Molecular spectroscopy offers solutions in a wide product range for accurate determination in food analysis. However, molecular spectroscopic methods have limited application in food analysis, therefore they must often be coupled necessarily with other analytical methods, such as chemometrics. Thus, this combination makes this technique a powerful tool in food quality analysis, including food characterization, classification, and authentication.

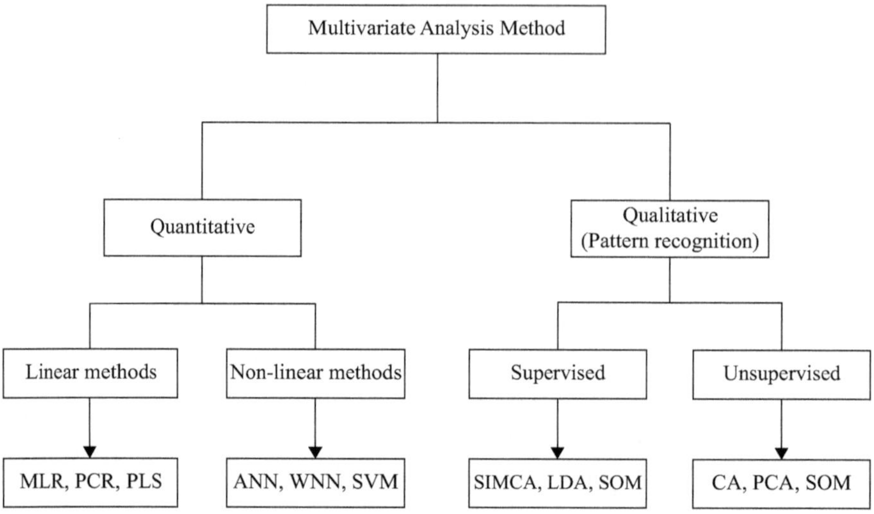

Figure 2.2. The multivariate analysis methods mainly applied to food analysis. MLR, multiple linear regression; PCR, principal component regression, PLS, partial least-squares; ANN, artificial neural networks, WNN, wavelet neural networks; SVM, support vector machine; SIMCA, soft independent modelling of class analogy; LDA, linear discriminant analysis; CA, Cluster Analysis; PCA, Principal Component Analysis; SOM, self organizing maps.

2.3 Chemometric methods

Recently, nearly all spectroscopic methods for food quality analysis employed the chemometric method for food analysis development and its validation. Therefore, this chapter would not be complete and interesting without a chemometrics section in food analysis. The reason for using the chemometric method in spectroscopic techniques is due to the fact that the spectroscopic spectra are large datasets that cannot be easily resolved by using conventional analysis. Thus, the chemometric methods are a critical method for solving large datasets of spectroscopic spectra for food analysis.

Chemometrics employs multivariate analysis to extract meaningful information from a complex analytical data set, such as in food analysis (Miller and Miller 2010). Using these methods, food analysis applies predominantly to regression and pattern recognition models (Callao and Ruisánchez 2018). In this context, multivariate analysis of food has enabled both qualitative and quantitative analyses. The most common multivariate analysis methods used in food analysis are given in figure 2.2. Here, for food qualitative analysis, such as classification, the methods used are mainly unsupervised and supervised learning methods (Esteki *et al* 2018). In unsupervised learning, methods split data sets into groups without specifying the sample categories, and use trends or patterns among samples. The sample is identified without prior information regarding groups or patterns. The most common unsupervised methods used in food analysis are cluster analysis (CA) and principal component analysis (PCA) (Berrueta *et al* 2007). Supervised learning methods apply a training step based on *a priori* categories of known object groups in order to build models that are subsequently employed to estimate unknown objects. The assumptions of supervised methods are the mutually independent target categories, and samples are assigned to groups to identify appropriate categories via correlations between classes and categories in order to construct classification models, which are later applied to put new samples to the most similar group for each case.

In food analysis, most of the methods begin with exploratory data analysis or unsupervised pattern recognition, followed with supervised learning steps (Baker and Inventado 2014). Linear discriminant analysis (LDA) is one of the most popular supervised models in food analyses, where the model is suitable to be employed for food characterization, authentication, and adulteration. In many cases, a model should be built to quantify some of the food properties or the extent of adulteration. This can be performed using linear or non-linear 'regression' models that afford accurate data fitting. The linear methods, e.g. multiple linear regression (MLR), principal component regression (PCR), partial least squares (PLS) regression, and orthogonal partial least squares (OPLS) regression, are compatible to first order data and fit for parallel factor analysis (PARAFAC). While N-way partial least squares (N-PLS) regression is more suitable with second-order data. The mainly applied non-linear regression methods in food analysis are artificial neural networks (ANN), wavelet neural networks (WNN) and support vector machine (SVM). In this context, non-linear models can be applied for food analysis by providing the input variables that have a non-linear relationship to the output variables (target characteristics).

By employing those aforementioned methods, a sample spectrum as the sample 'image', can be handled similarly to any data of the digital image analysis. For example, data pre-processing, segmentation, enhancement, feature extraction and data classification for the generation of applicable methods to spectra data use similar mathematical models to those applied in digital image processing for image analysis or pattern recognition. Chemometric methods resemble digital extraction and clean-up of the sample rather than using physical manipulation for clean-up, and this becomes the main drawback for the application of chemometric methods, so their calibration and validation requires a large number of samples. In addition, the methods are matrix-dependent, and thus every change in matrix composition needs a new calibration and validation process. Besides these drawbacks, the development spectroscopy combined with chemometric methods in food analysis is becoming more interesting nowadays (figure 2.3), and the further developments are being focussed toward the development of chemometric methods to be more robust and less affected by the food matrices.

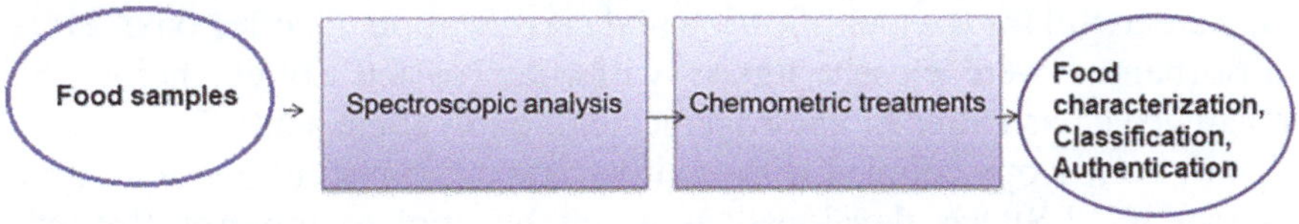

Figure 2.3. Application of spectroscopic analysis combined with chemometric methods for food analysis.

2.4 UV/vis spectroscopy

Among the spectroscopy techniques, there is spectroscopy based on the light absorption in the ultraviolet and visible region (UV–vis) (200–800 nm) and one of the most commonly employed techniques for food qualitative and quantitative analysis associated with the characterization of the organic and inorganic compounds in food matrices (Maria Porcu 2018). Thus, UV–vis spectroscopy is a commonly applied technique to analyze the quality of food products. It has become a very popular and important tool in different analytical areas worldwide owing to its availability, simplicity, flexibility, and versatility in various fields, including food analysis. In order to reduce sample and reagents volumes, especially for toxic solvents or very limited samples, a microvolume for the UV–vis spectrometric technique has been developed recently (Pena-Pereira *et al* 2011).

The coupling between UV–vis spectroscopy technique and chemometrics is mainly employed in the analyses of the food industry in order to develop quality control of foods and beverages, e.g. differentiation and variety of the origin (Palacios-Morillo *et al* 2013), adulteration detection and origin identification (Martins *et al* 2017), and others. Their development in the food industry quality laboratories has been very crucial as they satisfy not only customers but also both economic and public health issues as they allow the verification of various food products qualitatively and quantitatively.

2.4.1 Food components, pigments, and contaminants

Some recent applications of UV–vis spectroscopy combined with chemometric methods for the analysis of food components, pigments, and contaminants are presented. For example, UV–vis spectroscopy coupled with LDA has been developed for the characteristics verification of tea infusions prepared using boiling water (Diniz *et al* 2016). Here, 20 types of each tea (i.e. Brazilian black tea and green tea, Argentinean black tea, and green) were employed. One gram of each sample was prepared in distilled water (100 ml) at 90 °C. Quartz cuvettes were employed to get the spectra, where two absorption bands appeared in the region from 190 to 250 nm, and from 250 to 300 nm, as well as another wide absorption band found around 300–400 nm. These bands are in correspondence to phenolic compounds contained in tea infusions.

UV–vis spectroscopy and chemometrics have been applied for coffee sample classification, as pure normal coffee or pure peaberry (Suhandy and Yulia 2017). Here, PLS-DA has been used to analyze UV–vis spectral data for normal coffee and peaberry with excellent classification accuracy (100%) for both the training set and the test set. The loadings analysis for the spectral data (at 230, 250, 270, 310 and 350 nm) were useful for the identification and classification of coffee type. Moreover, their absorbances were closely linked with the roasted coffee major chemical components (such as caffeine, chlorogenic acid, and caffeic acid).

UV–vis spectroscopy coupled with multivariate curve resolution-alternating least squares (MCR-ALS) has developed as a reliable tool to monitor the oxidative stability of various edible oils in the UV–vis region (Gonçalves *et al* 2014). The oil samples from corn, soy, canola, sunflower, and olive were used. By using quartz cuvette, the spectra were acquired in the range of 300–540 nm. The heated sample spectra found a different peak maximum intensity value. This typical effect is common in the temperature variation cases, but not affected on MCR-ALS curve resolution. This method gave information related to oxidation products and degradation of tocopherol caused by the heating process.

UV–vis spectroscopy has been employed to determine pigments in extra virgin olive oil samples produced from olives harvested in different countries, and compared with HPLC-DAD (Lazzerini and Domenici 2017). The oil samples were measured in quartz cuvettes without any treatment, in the spectral range of 390–720 nm, with 1 nm resolution. The study showed that the ferritin B pigment can be used accurately to determine ferritin B by UV–vis spectroscopy, while the optimized HPLC-DAD method did not allowed one to determine it with sufficient accuracy.

UV–vis spectrophotometry has been developed to measured the mercury pre-concentration in water and fish samples using PAN (1-(2-pyridylazo)-2-naphtol) as colorimetric reagent (Fashi *et al* 2017). Here, the absorbance of Hg(II)–PAN complex was monitored at 554 nm, while the PAN maximum absorbance was monitored at 460 nm, with methanol used as the blank. The microvolume sample preparation procedure, related with the determination of Hg(II)–PAN complex microvolume, allowed a simple, rapid and accurate measurement of total Hg(II) ion in both water and fish samples.

2.4.2 Food adulteration and fraud

In the case of food characterization involving adulteration, it can be analyzed in two ways: employing qualitative analysis when the adulterant is unknown, and quantitative analysis, when the adulterant is known for satisfactory results. The UV–vis spectroscopy has been used in food and beverages quality control, and they have increasingly attracted much interest owing to their screening ability that allows the identification of the possible fraud that might be present in food products (Callao and Ruisánchez 2018).

UV–vis spectroscopy was applied for identification of counterfeit whiskey. This is important for a popular alcoholic beverage with high-economic value that is marketed worldwide (Martins *et al* 2017). In this case, using the original bottle, whiskey had been adulterated by mixing another lower quality whiskey. Combined with partial least squares for discriminant analysis (PLS-DA) it was able to discriminate the different whiskey brands and identify adulteration efficiency, such as mixing and dilution with another whiskey.

In food adulteration, one of the most crucial issues is related to consumer health risk or economic advantages (Johnson 2014). For example, olive oil has a higher commercial value than common oils due to high nutritional value and pleasant sensory characteristics. Therefore, it is often adulterated by the addition of other oils with lower quality, such as vegetable oils. UV–vis spectroscopy has also been applied in the adulteration of vegetable oils. For example, it has been developed to verify extra virgin olive oil adulteration with soy oil (Milanez *et al* 2017). UV–vis spectroscopy as a screening approach showed very promising results that can be used to support other current standard official methods that are used to identify oil adulteration.

Another application was to identify commercial pomegranate molasses adulteration with date molasses. Many different indicators that could be used to indicate adulteration were identified, such as anthocyanin concentration, polyphenol yield, total acidity, antiradical activity, and color intensity. In this context, UV–vis spectroscopy was developed as the screening tool to determine adulteration along with other methods, i.e. ATF-FTIR and HPLC (El Darra *et al* 2017). Here, samples used were two authentic natural pomegranate molasses, three commercial pomegranate molasses, and one natural date syrup. The absorption bands were found in the spectral range from 190 to 1100 nm in the UV–vis region with quartz cuvettes. The natural pomegranate molasses tested showed peak amplitudes at higher values with similar maximum optical density values, while the commercial pomegranate molasses and date syrup presented peak amplitude with lower values. This phenomenon may be associated with diversity and different concentration of the phenolic compounds as absorbent molecules.

Coffee is one of the most popular drinks and beverages worldwide, owing to its benefits to health, including stimulant property and intrinsic characteristics. Coffee type variety is relatively large. In this context, UV–vis spectroscopy combined with LDA has been developed to identify coffee extracts irregularities (Barbosa *et al* 2015). Here, the method has benefits over other standard analysis employed for this type of analysis as it provides a rapid and simple analysis for the coffee extracts. Moreover, it also offers consumer safety including with regulatory agencies to avoid fraudulent product labeling.

The UV–vis spectroscopy has also been developed to determine *Coffea canephora* varieties, the robusta addition to *Coffea arabica*, and the complementarity of data produced with another technique, i.e. synchronous fluorescence spectroscopy (Dankowska *et al* 2017). Here, coffee extracts were prepared with distilled water at 95 °C at 6% (w/v), then filtered and diluted in the 1:120 ratio (v/v). A sum of 147 spectra were recorded. All of the measurement was done in triplicate for genuine coffee, and in duplicate for the coffee blends samples. The spectra were measured to allow the UV–vis spectroscopy to discriminate coffee robusta, *Coffea arabica*, and their blends. The results presented that the aqueous extracts of *Coffea arabica* and *Coffea canephora* varieties showed their UV–vis spectra had significant differences. These studies showed the UV–vis spectroscopy gathered numerous and important methods employed in the food products quality control improvement and development.

2.5 Fluorescence spectroscopy

Fluorescence is the emission of light after UV or visible light absorption of a fluorescent molecule (fluorophore). Common fluorophores are fluorescein, acridine orange, quinine, pyridine 1 and rhodamine B (Lakowicz 1999). The fluorescence high specificity is owing to the two spectra used (excitation and emission spectra), while its high sensitivity is due to emission radiation measurement versus absolute darkness. On the other hand, the wavelengths of excitation and emission used can limit this technique's ability for food quality determination. Therefore, the application of synchronous fluorescence spectroscopy (SFS) could overcome this limitation (Karoui and Blecker 2011). This can be achieved by the excitation and emission wavelengths variation that enables simultaneous determination of food compositions in various samples.

Fluorescence spectroscopy allows a rapid, sensitive, and selective analytical method to obtain food composition information, such as nutrition, function, and composition that can be employed as a food fingerprint, such as e.g. fish, meats, dairy products, edible oils, wines, etc. This technique used for food characterization and authentication is increasing currently, due to its potential being enhanced by chemometric methods (Esteki *et al* 2018). This technique combined with chemometrics applications in food analysis has been reviewed recently. In this review, using different chemometrics methods, such as PCA, CA, PLS, LDA, and PARAFAC in fluorescence spectroscopy, they have applied food characterization and classification, such as meat, fish, egg, dairy products, wine, honey, and edible oils.

2.5.1 Food compositions, additives, and contaminants

In food analysis, the high specificity and sensitivity of fluorescence spectroscopy have been exploited for determination of food compositions, additives, and contaminants. In this context, this technique is mainly linked with other techniques, such as LC/GC (liquid chromatography or gas chromatography), where the detector is a fluorimeter. The combination of these techniques has benefits for detection of protein structural changes, carbohydrates analysis and analysis of lipids in oils (Nakai and Horimoto 2000). Other important applications are to determine food additives (e.g. aspartame, salicylates), and trace concentrations of food contaminants, such as pathogenic bacteria (*Escherichia coli*, *Salmonella*), toxins (mycotoxins), antibiotics (e.g. tetracycline, oxytetracycline, penicillin).

In food composition analysis, fluorescence spectroscopy has been employed in analysis of cereals and cereal products (Zandomeneghi 1999). It has been used to differentiate between different cereal flours (e.g. rice, maize). A clear difference was found between samples of the two groups by using red and white wheat kernels emission spectra (Ram *et al* 2004). Here, this difference has been associated with the morphological variation in the two wheat varieties. Furthermore, by using the emission spectra of ferulic acid and riboflavin, fluorescence spectroscopy has been successfully applied to determine the efficiency of wheat flour refinement and milling (Symons and Dexter 1991).

Fluorescence spectroscopy coupled with pattern recognition was applied to classify Argentinean white wines related to grape variety based on excitation–emission matrices (Azcarate *et al* 2015). In this case, unsupervised pattern recognition was employed using PCA and PARAFAC. While SIMCA, U-PLS-DA, N-PLS-DA, and SPA-LDA models were applied for classification and specification of the grape type in each wine as the test set. The results found that UPLS-DA and SPA-LDA with 76% and 80% sensitivity, respectively, gave very promising results for the classification of the wines.

In food contamination, fluorescence spectroscopy can be said to be one of the most valuable tools for the determination of food poisoning causes by analyzing the concentration of the toxin, especially mycotoxins. This is due to nearly all mycotoxins, besides aflatoxins (aflatoxins B1 and B2) showing blue fluorescence, while aflatoxins G1 and G2 exhibit yellow–green fluorescence (Li *et al* 2009), and the latter did not show fluorescence. This is due to this technique being combined with other analytical techniques, where the detector used a fluorimeter.

Capillary electrophoresis (CE) with laser-induced fluorescence as the detector has been used to determine Ochratoxin A in roasted coffee, corn, and sorghum (Corneli and Maragos 1998). Similarly, fumonisin B1 in corn samples has been determined using CE that was labeled with fluorescein owing to a rare UV chromophore (Maragos 1997). The fluorescence polarization (FP) immunoassay has been used as a rapid test for deoxynivalenol (DON) determination in wheat (Maragos and Plattner 2002). The principle of FP immunoassay is based on a toxin-specific antibody interaction with a toxin–fluorophore conjugate (tracer) to effectively decrease the tracer rotation rate. Here, the test utilizes the competition between DON and a novel DON–fluorescein tracer for a DON-specific monoclonal antibody in solution. The antibody binding to the tracer makes polarization increase, with the presence of the free toxin, then less antibody is bound to the tracer, which in turn, causes polarization to be reduced. Fumonisin mycotoxin detection has also been developed using FP immunoassays (Maragos *et al* 2001).

2.5.2 Food authentication, origin, and adulteration

In food authentication or adulteration, olive oil has been paid much attention. This is due to olive oil being one of the high-value and most important edible oils where the price is based on its quality. Extra virgin olive oil is the most expensive oil owing to its very high quality. It is often adulterated for economic reasons, by the addition of cheaper oils (e.g. refined olive oil, synthetic olive oil–glycerol products, residue oil, seed oils, and nut oils). Therefore, a simple and rapid tool to determine this adulteration is needed for labeling purposes and quality control of this product (Karoui and Blecker 2011).

Fluorescence spectroscopy has been developed for adulteration determination of hazelnut oil in virgin olive oils (Sayago *et al* 2004). In this technique, the excitation wavelength at 360 nm was employed to discriminate between olive oil and other vegetable oils (i.e. refined olive oil, olive residual oil, corn oil, sunflower oil, soybean oil, and cotton oil) (Kyriakidis and Skarkalis 2000). The oil samples tested, all exhibited a strong fluorescence band (at 430–450 nm) but not for virgin olive oil that showed a low intensity (at both 440 and 455 nm), and a medium band around 681 nm with a strong one at 525 nm. The latter two bands were associated with chlorophyll and vitamin E, respectively. All refined oils exhibited only one intense peak (at 445 nm) is owing to fatty acid oxidation formed by the large polyunsaturated fatty acids present in these oils.

Fluorescence spectroscopy combined with three-way calibration methods has been developed for detection of olive oils adulteration from the protected designation of origin (PDO) 'Siurana' with olive pomace oil (Guimet *et al* 2005). The method employed unfold-PCA and PARAFAC for exploratory data analysis, afterword Hoteling T2 and Q statistics have been applied for rapid screening for adulterated samples. The result shows that the differentiation of pure (non-adulterated) and adulterated samples was increased by employing LDA (100% sensitivity). While N-PLS regression was employed for quantitative adulteration (LODs of 5%).

Fluorescence spectroscopy combined with digital imaging has been evaluated and developed for detection of extra virgin olive oil adulteration (Milanez *et al* 2017). Here, in order to predict an adjustment function and use it for samples tested by the secondary instrument, the direct standardization (DS) and piecewise direct standardization (PDS) methods were employed to transfer the sample data sets. While for comparison, SPA–LDA and PLSeDA were used in this case. In order to test the standardization procedures there was a requirement for efficiency in constructing multivariate classification models based on digital images and fluorescence spectral data. Before standardization, SPA-LDA and PLS-DA models reached the same accuracy, i.e. 47% with the histograms of the digital images 54% and with fluorescence emission data. Here, standardization improved sensitivity substantially, i.e. 82% for the digital histograms and 88% for the fluorescence emission spectra.

Fluorescence spectroscopy and chemometrics have been applied for the detection of extra virgin olive oil adulteration with other types of olive oil (Durán Merás *et al* 2018). Front-face fluorescence spectroscopy was used to obtain excitation–emission matrices (EEMs) for extra virgin olive oils and samples adulterated with olive or pomace oil. PARAFAC as an unsupervised method and LDA–PARAFAC, as well as discriminant, DA-UPLS as supervised pattern recognition methods were used to analyze fluorescence images obtained. The LDA–PARAFAC performance to differentiate between classes was evaluated. Three-dimensional distributions of canonical vectors used for determination of the separation level among the classes resulted. DA–UPLS discriminating performance was evaluated from predicted value plots for calibration and validation samples. The models are able to detect extra virgin olive oil adulteration of about 15% of olive oil and 3% olive pomace oil. Moreover, UPLS regression was employed for the adulterants quantitative determination of the extra virgin olive oil adulteration level.

Fluorescence spectroscopy can be used as a tool that allows for the excellent models for oligomers, vitamin E, and anisidine and iodine values in deteriorated oil after repeated frying cycles that can be used for deterioration evaluation of frying oil (Engelsen 1997). Total luminescence and synchronous scanning fluorescence spectroscopy have been developed for edible oils characterization and differentiation, such as sunflower, soybean, peanut, rapeseed, olive, linseed, corn oils, and grape seed (Sikorska *et al* 2005). All oils show a strong peak of the luminescence spectra of the excitation spectrum at 290 nm and in the emission spectrum at 320 nm. These peaks corresponded to tocopherols. Moreover, some oils showed a long-wavelength peak at 405 nm in excitation and 670 nm in emission, where they associated with the chlorophyll group pigments. Other additional bands found in the fluorescence spectra associated with unidentified compounds. The synchronous-scanning fluorescence spectra also had similar bands for chlorophylls, tocopherols and unidentified components, where KNN and LDA were applied for the oils classification. The overall results showed the fluorescence techniques' ability to discriminate and characterize vegetable oils tested. Furthermore, the synchronous front-face fluorescence spectroscopy has been developed to discriminate used frying oil (UFO) from edible vegetable oil (EVO), predicting the UFO usage time, and detecting EVO adulteration with UFO (Tan *et al* 2017). Here, PLS-R was applied to build a model for high accuracy prediction of the UFO heating time, and EVO adulteration with UFO.

Fluorescence spectroscopy coupled with chemometrics has been developed for cheese adulteration detection with vegetable oils (Dankowska *et al* 2015). Here, MLR models were used to determine the adulteration level based on the spectral data. The adulteration LOD was 3.0%. While SPA–LDA and PCA–LDA were also applied for better classification accuracy. Furthermore, they have been used for determination and quantification of adulterated butter with palm and coconut oils (Dankowska *et al* 2014). The fluorescence spectra were in the range of 240–700 nm (with 10, 30, 60 and 80 nm intervals), where 60 nm gave the LOD of adulteration at 5.5%. In other work, front-face fluorescence spectroscopy has been applied for unifloral and multifloral honey types authentication, which previously were classified based on traditional methods, e.g. pollen, chemical and sensory analysis (Ruoff *et al* 2005). Here, PCA and LDA were used for spectral data for effective authenticating of the honey botanical origin and analyzing geographical origin within the same honey.

The used fluorescence combined with UV–vis spectral data has been employed for Chinese canned lager beer samples classification (Tan *et al* 2015). Right-angle synchronous fluorescence spectra (SFS) and visible spectra in the range 380–700 nm obtained at three different intervals (30, 60 and 80 nm) for undiluted beers along with UV spectra in the range 240–400 nm were used in this study. The LDA model used was capable of classification sensitivity between 78.5%–86.7%. However, the individual spectroscopies had accuracy only in the range of 42.2%–70.4%. This data shows that the synergistic effect of combined fluorescence with UV–vis spectroscopies could increase classification accuracy.

2.6 IR spectroscopy

Infrared (IR) radiation is electromagnetic radiation in the range of 780 nm–1 mm. The IR range can be divided into the three regions, i.e. NIR (near-infrared) in the range 780 nm–5 μm, MIR (mid-infrared) within 5–30 μm, and FIR (far-infrared) within 30–1000 μm (Stuart 2004). Therefore, IRS (infrared spectroscopy) is a technique based on the vibrations of the atoms of a molecule and has been employed for analysis of food quality and authenticity. IRS has become an interesting and flexible tool based on its simple operation, and it needs less sample preparation (Abbas *et al* 2012). Moreover, its high sensitivity and specificity has allowed IRS to be employed as a fingerprint generator. However, the NIR and MIR regions are commonly employed for food analysis. This is due to FIR (below 400 cm^{-1} down to 10 cm^{-1}) and particularly the region below 200 cm^{-1} not having many useful spectra–structure correlations for food compounds, hence it is not readily useful in food analysis. Moreover, the FIR region is only absorbed by compounds that have halogen atoms, inorganic compounds, and organometallic compounds, where torsional vibrations and stretching modes of hydrogen bond appear in this region (Shurvell 2006).

Generally, an IR spectrum is created by passing IR radiation via a sample and detecting a kind of fraction that has absorbed the incident radiation at a particular energy. This energy in an absorption spectrum appears as any peak links to the vibration frequency of a sample molecule. Hence different types of vibrations found in the IR spectra give information on molecular structure via the normal modes of the molecule vibration frequencies. NIRS is based on the electromagnetic radiation absorption at wavelengths ranging from 780–2500 nm, where its peaks appear from molecular overtones and vibrational modes combinations (Stuart 2004), while MIRS (4000–400 cm^{-1}) is based on real-time detection of the functional group vibrations and their rotational–vibrational effects. NIR spectra are obtained from composite superposition, overtone and high-frequency absorption bands at shorter wavelengths (Stuart 2004). While MIR spectra at fingerprint range are the result of absorbed IR radiation caused by atoms' fundamental vibrations in the molecules (Shurvell 2006).

IRS provides simple and low-cost tools for screening of samples in preliminary routine food analyses. NIRS particularly is a promising method for its accuracy, sensitivity, and ability to produce spectra for solid and liquid samples with no required for prior manipulation (Blanco and Villarroya 2002). Hence, the IRS methods are among the most frequently and commonly used spectroscopy methods in food analysis. The application of FT–NIR, MIR including Raman spectroscopy for characterization of food products has been reviewed for food authenticity and adulteration (Lohumi *et al* 2015). NIR and MIR combined with chemometric methods mainly have been developed for a tool for simple and rapid methods of food analysis (e.g. quality control, authenticity, and adulteration) in various food products as shown in table 2.1.

The FTIR applications for authentication and potential harmful adulterants detection based on various chemometric methods were reviewed (Rodriguez-Saona and Allendorf 2011). The detection of adulteration in food, (honey, orange juice, alcoholic beverages, and milk powder) based on NIR spectra data coupled with PLSR was reported. In addition, the MIR applications in food adulteration detection mainly employing PLSR and SIMCA were reported, such as extra virgin olive oil, honey, juice, butter, and wine as well as lard in cake and chocolate.

Table 2.1. Current applications of IR spectroscopies combined with chemometric methods for food analysis.

Food sample	IRS technique	Food analysis objective	References
EVOO	IR	EVOO origin tracing	
EVOO	FTIR	EVOO authentication	
EVOO	FTIR	The genetic variety prediction EVOO produced in the region of Valencia, Spain	
EVOO	FTIR	Adulteration detection of EVOO with peanut oil	
EVOO	FTIR	Distinguishing the geographic origin of EVOO	
EVOO	FTIR	Confirmation of EVOO from Liguria	
EVOO	MIR	Rapid adulteration detection of EVOO	Georgouli *et al* (2017)
Olive oil	MIR	Differentiation of monovarietal olive oils mixtures	Gurdeniz *et al* (2007)
Olive oil	NIR	Analyzing olive oil	
Almond oil	FTIR	Authenticate cultivars based on the volatile compounds	
Oils	FTIR	Functional food oils authentication	
Edible oils and fats	FTIRFT-NIR	Edible oils and fats discriminant analysis	Yang *et al* (2005)
Moroccan olive cultivars	ATR-FTIR	Moroccan olive cultivars classification	
Honey	NIR	Adulterants detection: sweeteners in honey	
Honey	FTIR	Rapid detection of honey adulteration with invert cane sugar	
Roasted and ground coffee	FTIR	Determination of many adulterants in roasted and ground coffee	
Roasted coffee	FTIR	Defective and non-defective roasted coffee discrimination	
Roasted coffee	ATR-FTIR	Simultaneous multiple adulterants determination in ground roasted coffee	Reis *et al* (2017)
Red wines	NIR	Spanish and Australian Tempranillo red wines geographic classification	

Maple syrup	IR	Adulterants discrimination and classification in maple syrup	Paradkar *et al* (2003)
Orange juice	IR	Adulteration detection in freshly squeezed orange juice	
Fish fillet	FT-NIR, FT-IR	Fsh fillet authentication	Alamprese and Casiraghi (2015)
Meat products	MIR	adulteration detection in uncooked and cooked meat products	
Minced beef	NIR, MIR	Minced beef adulteration detection with turkey meat	Alamprese *et al* (2013)
Beef meatballs	NIR	Adulteration detection of Pork in beef meatballs	Kuswandi *et al* (2015)
Beef	NIR	Adulterants detection and quantification in fresh and frozen-thawed minced beef	Morsy and Sun (2013)
Sweet potatoes	NIR	Discrimination of adulterated pure, powdered, purple sweet potatoes with the sweet potato flour	Ding *et al* (2015)
Foods	FTIR	Rapid adulteration and authentication of food	Rodriguez-Saona and Allendorf (2011)

Note: EVOO, Extra virgin olive oil.

2.6.1 NIR spectroscopy

NIRS has been mainly used in food quantitative analysis, such as to determine carbohydrate, protein, fat, and moisture content in various foods. Table 2.2 describes principal absorption bands of protein, starch, oil, and water that are found in the NIR region (Abbas *et al* 2013). These bands are mainly due to vibrations that involve C–H, O–H, and N–H. Moreover, it might be due to S–H and C=O bonds that are responsible for the observed absorption bands majority found in the NIR region.

The greatest advantage of NIRS is to allow one to determine several food components simultaneously in a short time. The precision of NIRS in various food analyses is comparable, and often better than conventional methods, such as chemical or biological analysis. But NIRS has drawbacks in quantitative analysis of food, e.g. it needs calibration using known composition samples. This drawback has limited the application of NIRS as it needs a long time and high budget to build reliable and valid calibrations. This drawback is exacerbated by the calibration instability problem that is caused by food sample or instrument characteristics changing over time, making it necessary to create frequent recalibration. Moreover, due to optical differences in NIRS between instruments, it is not possible for calibrations transferability. There are other NIRS drawbacks in food analysis, e.g. the high-precision instruments required, the data treatment complexity, and the slow sensitivity for minor food elements (Li-Chan *et al* 2006).

Table 2.2. Principal absorption bands observed in the NIR region of food major components, such as protein, starch, oil, and water.

Components	Wavelength (nm)	Assignment
Proteins	1208	2nd overtone C–H stretching
	1465	1st overtone H–N and O–H stretching
	1734	1st overtone C–H stretching
	1932	Combinations N–H and O–H stretching
	2058	
	2180	
	2302	Combination C–H stretching
	2342	
Starch	1204	2nd overtone C–H stretching
	1464	1st overtone H–N and O–H stretching
	1932	Combinations N–H and O–H stretching
	2100	
	2290	Combination C–H stretching
	2324	
Oil	1210	2nd overtone C–H stretching
	1406	1st overtone H–N and O–H stretching
	1718	1st overtone C–H stretching
	1760	
	2114	Combinations N–H and O–H stretching
	2308	Combination C–H stretching
	2346	
Water	1454	1st overtone O–H stretching
	1932	O–H combinations

2.6.1.1 Applications

The application of NIRS in food analysis, such as for wheat and wheat products, include detection of water absorption, damaged starch, flour yield, dough development time, volume and extensibility of loaf, etc. NIRS application for determination of the protein and moisture in flour and wheat is routine work in many flour mills nowadays. It is applied for testing each wheat delivery to make decisions regarding price, acceptance, and binning, including for measuring hardness conditioning time, and to check the flour fulfils the specifications required prior to sending to the customer (Osborne 2000).

Currently, NIRS is often coupled with a hyperspectral imaging system (NIR–HSI). They have been applied for the determination of Canadian wheat classes (Mahesh *et al* 2008). They used 75 relative reflectance intensities that were taken from the scanned images for wheat classes differentiation employing chemometrics (statistical classifier and ANN) with excellent classification accuracy (100%) of wheat classes. The HSI has also been applied to determine spectral differences between healthy and damaged products, that are related to these products' chemical composition. In addition, insect-damaged kernels have been studied (Singh *et al* 2009, 2010), where it was found that insect-damaged kernels had less starch compared to healthy products, as the insects had consumed starch during the products' development.

NIRS coupled with latent variable models has been developed based on their spectral data to predict proximate chemical composition, fatty acid, fillet yield and cooking loss, as well as to classify between raw freeze-dried rainbow trout fillets and cooked freeze-dried fillets using the available dataset by the rearing farm and genetic strain of rainbow trout (Dalle Zotte *et al* 2014). Here, the result shows no major differences were observed between them, except for the prediction of some chemical elements of interest, such as polyunsaturated fatty acid, where it was better predicted in cooked freeze-dried fillets. The food authenticity study involves establishing whether a sample is original as its description, such as product purity, geographical origin, etc. NIRS application for food products authenticities, such as fruit pulps, coffee, milk powders, pig carcasses, orange juice, rice, sausages, vegetable oils, sugars, wheat grain, and wheat flour have been reviewed in the literature (Downey 1996). In this case, NIRS is needed to classify within the possible class series, in order to quantify adulteration or to identify a particular kind of adulteration.

NIR spectral data and chemometric method have been applied to determine adulteration of pork in beef meatballs (Kuswandi *et al* 2015). Here, LDA is for qualitative analysis, while PLS regression is employed for quantitative analysis. Both the training set and the test set were obtained from first-derivative spectra, the resulting LDA models could be employed accurately to classify adulteration of pork in beef meatball samples. Moreover, the PLS and LDA results were in good agreement with the reference method using the immunochromatographic assay. The method demonstrated a rapid, efficient tool for detection of pork adulteration in beef meatballs that can also be applied as halal authentication.

NIRS with LDA and PLSR have been evaluated as a non-destructive and rapid tool for detection and quantification of various adulterants in minced beef, i.e. fresh and frozen-thawed (Morsy and Sun 2013). PLS regression was performed using cross-validation and applied with different parameters (e.g. standard error of prediction/SEP and determination coefficients/R^2) for the adulteration of pork in minced beef samples, offal, and fat trimming. PLS–DA and LDA models were developed using the most suitable wavelengths for differentiating unadulterated/pure from the adulterated samples. Both models show a high classification specificity and sensitivity, and testified to the vis–NIR spectroscopy applicability for pork adulteration detection in minced beef.

NIRS coupled with multivariate analysis has been applied to discriminate pure canola oil from samples adulterated with palm oil (Hussain *et al* 2014). Using LDA, the method was able to detect adulterated samples with 100% accuracy, with LOD of 3.23%. In addition, NIRS combined with chemometrics has been applied to identify and determine powdered, pure and adulterated sweet potato (Ding *et al* 2015). PCA with the NIR spectral data enabled many potato type samples to be distinguished, and also total antioxidant activity (TAA) and total anthocyanin (TA) could be determined. Using LDA and KNN, they could accurately discriminate among adulterated samples, white sweet potato, and purple sweet potato.

2.6.1.2 Combined NIR and MIR

Recently, a methodology based on the NIR and MIR measurements combined with chemometric data processing was reported (Lin *et al* 2014). This combined method could tackle the drawbacks of both spectroscopy techniques and offer more reliable and accurate results in food analysis. For example, the potential of FTIR combined with NIRS for detection and discriminating adulterants, e.g. cane and beet invert syrups, and cane and beet sugar solutions, in maple syrup, was investigated (Paradkar *et al* 2003). The NIR region was used to analyze carbohydrates, while the FTIR spectra were used for carboxylic acids, carbohydrates and amino acids analysis, along with the NIR spectra (from 1100 to 1660 nm). Here, LDA and canonical variate analysis (CVA) were suitable for classification, while PLS and PCR enabled quantification analysis. FTIR showed the most accurate selection for determination of adulterants (e.g. pure beet and cane sugar solutions) in maple syrup.

NIR and MIR combined with LDA were applied for adulteration detection of minced beef with turkey meat. Pure bovine and pure turkey minced meat samples, and minced beef adulterated with turkey meat in proportion mixtures from 5% to 50% (w/w) were prepared for this purpose (Alamprese *et al* 2013). These combined NIR and MIR techniques allowed for rapid and highly accurate identification of species in minced meat products, even with low selectivity. Here, LDA was employed for identifying adulteration classification and PLS for adulteration quantification level.

2.6.2 MIR spectroscopy

In food analysis, MIRS can give information on structure–functionality relationships of food compounds and can be used as a tool for food quantitative analysis. Therefore, MIRS is a highly valuable technique for both food-related study and food product quality control. The important MIR bands related to the food major components (proteins, fats, carbohydrates, and water) are given in table 2.3 (Stuart 2004). Here, the 'fingerprint' bands are linked mainly to bending and skeletal vibrations, while the X–H stretching region fundamental vibrations are commonly due to C–H, O–H and N–H stretching. C≡C and C≡N bonds vibrations are mostly found in the triple-bond stretching region. While absorption bands relating to C=C, C=O, and C=N occur in the double-bond region.

The Fourier-transform infrared spectroscopy (FTIR) is mainly employed in food analysis. Similarly, with NIRS, MIRS is employing the analysis of protein, fat, carbohydrate, and moisture content in food products. Due to the water-intense absorption of MIR radiation, therefore, FTIR was applied to determine emulsions moisture of food, such as butter (van de Voort *et al* 1992) and mayonnaise (van de

Table 2.3. The major bands of food components (proteins, fats, carbohydrates, and water) localized in the MIR region.

Food components	Wavenumber (cm^{-1})	Assigment
Proteins	1600–1690	Amide I (C=O stretching)
	1480–1575	Amide II (C–N stretching and N–H bending)
	1230–1300	Amide III (C–N stretching and N–H bending)
Fats	2800–3000	C–H stretching
	1725–1745	C=O stretchnig
	970	C=C–H bending
Carbohydrates	2800–3000	C–H stretching
	800–1400	Skeletal stretching and bending
Water	3200–3600	O–H stretching
	1650	H–OH stretching

Voort *et al* 1993). FTIR is a powerful tool for the investigation of protein secondary structure, based on the amide I region examination (1600–1700 cm^{-1}) (Mejri *et al* 2005). Proteins are mainly applied as ingredients in the food processing industry due to their properties (e.g. gelation, emulsification, and thickening). The properties are highly linked to the secondary protein structure that could change during food processing and storage (Li-Chan *et al* 2006).

2.6.2.1 Food components analysis

Currently, MIRS plays an important role in the food chemistry study on fats and oils. For instance, an important investigation result in this field of an absorption band at 996 cm^{-1} has been identified as associated with isolated trans double bonds characteristic. These double bonds contained in oils and fats mainly have the cis configuration, therefore cis–trans isomerization extensively takes place during the industrial processes of catalytic hydrogenation that are mostly used to convert oils to fats and to improve the polyunsaturated oils' oxidative stability. Another field of FTIRS application that has been studied is its application in the oil quality and stability analysis. This is the general cause of fats and oils deterioration, as well as of any lipid-containing food, caused by a reaction between atmospheric oxygen and unsaturated lipids under ambient conditions, namely lipid autooxidation. This autoxidation causes oxidative rancidity that leads to off-flavors and unpalatable odors. FTIRS showed itself to be the most accurate and direct monitoring method of gross changes over time of frying oil (Engelsen 1997, Moros *et al* 2009).

FTIRS has also been developed as a quantitative method that was applied to the oxidative state of frying oils monitoring (Dubois *et al* 1996). This was performed based on the anisidine value (AV) determination, where there is a measurement of aldehydes as the main secondary oxidation products in polyunsaturated oils. Moreover, FTIRS has also been used as an alternative tool to measure the peroxide value (PV) test that is commonly used in the oils and fats industry to measure the refined oils' oxidative status and stability. The FTIR method is based on the measurement of the hydroperoxide OO–H stretching absorption that is found at 3444 cm^{-1} in the neat oils spectra (van de Voort *et al* 1994). It was employed to study in more detail the thermal oil degradation mechanisms (i.e. thermally induced oxidative processes) in extra virgin olive oils (Navarra *et al* 2011). Moreover, it was used for determination of other minor components present in oils, such as free fatty acids in refined (Bertran *et al* 1999) and crude oils (Che Man *et al* 1999), phospholipids in vegetable oils (Nzai and Proctor 1998), β-carotene in palm oil (Moh *et al* 1999) etc.

2.6.2.2 Food authentication and adulteration

FTIRS has been developed as a simple and rapid tool for detection of redmullet and plaice substitution with Atlantic mullet and flounder (Alamprese and Casiraghi 2015). Here, it can also be used to discriminate and identify fresh and frozen–thawed fillets of Atlantic mullet. LDA was used to differentiate Atlantic mullet from more valuable red mullet with high specificity and sensitivity.

MIRS is also employed as a tool to investigate olive oil adulteration by other edible oils since these oils are commonly used as adulterants in virgin olive oil. Another adulterants are lower quality olive oil, such as refined or pomace olive oil, or other vegetable or seed oils (e.g. corn, sunflower, soybean, cottonseed, peanut, and poppy seed oils, Harwood and Aparicio 2000). MIRS coupled with chemometrics has been used as a rapid tool to identify and quantify extra virgin olive oil (EVOO) adulteration with vegetable oils (cottonseed, rapeseed, and corn–sunflower binary mixture) (Gurdeniz and Ozen 2009). EVOO adulteration by different palm oil concentrations was investigated by using FTIRS (Rohman and Man 2010). In addition, MIRS and PLS–DA have been developed to identify and quantify EVOO adulterated with different edible oils. The MIR spectral data were processed to wavelet compression before PCA. The LOD of about 5% for diagnosing adulteration was achieved for corn and sunflower oil binary mixtures, including cottonseed and rapeseed oil. Here, the data were analyzed using PLS–DA as a model for fraud detection irrespective of the adulterant oil types. The FTIR–ATR based method combined with chemometrics (PCR, PLS–R or LDA) has been developed to screen EVOO adulteration with peanut oil (PO) (Vasconcelos *et al* 2015). PCA was used to specify the main frequencies responsible for the biochemical differences among several edible oiltypes, while LDA was used successfully for EVOO binary classification and adulterated samples containing PO 0.5% (v/v). Furthermore, LOD was predicted to be about 10% (Gurdeniz *et al* 2007). It has also been found that not only is EVOO adulterated, but also virgin coconut oil (VCO), by low-quality edible oil that possesses several biological activities (e.g. antiviral and antimicrobial activities). FTIRS combined with chemometrics has been used to detect the adulteration VCO level with sunflower and corn oils (Rohman and Che Man 2011).

The attenuated total reflectance (ATR) MIRS application was developed as a rapid tool for milk adulteration identification and quantification (Santos *et al* 2013). Here, the milk samples were spiked with different adulterant concentrations (i.e. whey, synthetic urine, hydrogen peroxide, urea, and synthetic milk). The SIMCA as pattern recognition analysis and PLSR were applied for prediction of adulterant levels. Results proved that MIRS could serve as an alternative method for screening of potential fraudulent cow's milk adulteration in the dairy industry. In addition, MIRS coupled with LDA has been developed successfully to distinguish between pure beef and beef adulterated with heart, kidney, liver, or tripe with high specificity and sensitivity (Al-Jowder *et al* 1999). In addition, MIRS and PLS-DA have been proposed as a novel monitoring tool for adulteration detection of ground roasted coffee with spent coffee grounds, roasted corn, roasted coffee husks, and roasted barley (Reis *et al* 2017). Here, ATR and DR were compared for their performances. By applying PLS–DA to the spectral data, in this case, it found that DR reduced the training and test sets misclassification.

2.6.2.3 Combined FTIR–Raman

These combined methods between two or more spectroscopies could offer more reliable and accurate results in food analysis since the drawbacks of both spectroscopy techniques could be eliminated. The FTIR and Raman fingerprinting method coupled with a novel dimension reduction method has been proposed for detection adulterants in EVOO by hazelnut oil (Georgouli *et al* 2017). Continuous locality preserving projections (CLPP) has been applied for the continuous structure of the data manifold to be preserved in order to improve the performance classification and accuracy of adulterant detection. The combined method specificity and sensitivity were against Raman and FT-IR spectroscopies coupled with kNN (k-nearest neighbor) and LDA, where combined method shows better accuracy and sensitivity in detecting adulterants.

2.7 Raman spectroscopy

Raman spectroscopy (RS) is a non-destructive analytical technique using Raman scattering of light from molecules that causes shifted energy frequencies. Raman signals are obtained from light inelastic scattering of samples. Therefore, the resulting frequency shift is consisting of information regarding the vibrational modes involved (Larkin 2011, Craig *et al* 2013). RS can be developed as FTRS (Fourier transform Raman spectroscopy), dispersive Raman spectroscopy, SORS (spatially offset Raman spectroscopy) and SERS (surface-enhanced Raman spectroscopy). The Raman spectra can show a particular compound fingerprint that is useful for analyzing elements contained in different food samples, and they also give the basic information for structural and qualitative analyses (Yang and Ying 2011).

RS and MIRS are complementary techniques, but RS has some benefits over MIRS. For example in Raman spectra, water has weak RS absorption so that it does not cause interference. Therefore, no interference from water is necessary for the method reliability in determination of food authenticity since, in fact, most foods contain high or low water. Compared to IR spectral bands, the Raman spectra absorption bands are narrower. This is crucially important for RS application, due to narrow bands making less overlapping, especially in foods with complex matrices.

Table 2.4. Raman bands of major food components (carbohydrates, proteins, fats, and water).

Food components	Wavenumber (cm^{-1})	Assignment
Carbohydrates	836	C–C stretching
	1064	C–O stretching
	2912	C–H stretching
	2944	
	3451	O–H stretching
Proteins	510	S–S stretching
	525	
	545	
	630–670	C–S stretching
	700–745	
	1235–1245	Amide III (C–N stretching and N–H bending)
	1600–1700	Amide I (C=O stretching and N–H bending)
	2550–2580	S–H stretching
	2800–3000	C–H stretchnig
Fats	1441	CH_2 bending
	1457	CH_3–CH_2 bending
	1656	C=C stretching
	2855–2960	C–H stretching
Water	3200–3600	O–H stretching

RS is employed to investigate carbohydrate, protein and fat structure as well as water in food samples. Table 2.4 presents Raman bands of the main components in food products (Herrero 2008).

In food analysis, RS also can be used directly for transparent food packaging materials, such as plastic and glass bottles, wrappers, bags, etc (Ellis *et al* 2015). Since inelastic scattering in RS is weak, improved techniques like SERS, allow lower detection limits, and make potential applications in the detection of food adulterations (Danezis *et al* 2016). For instance, three Raman modalities (normal Raman, FT–Raman, and SERS) coupled with the chemometric method (PCA) were developed as a tool for detecting Sudan I dye in culinary spices. The results prove that SERS is the most suitable tool for obtaining a proper Raman signal for food analysis as a complex matrix (Di Anibal *et al* 2012).

2.7.1 Food components analysis

Similar to IRS, RS is employed in food analysis, for protein, carbohydrate, fat, and water in food samples. As described above, in water weak Raman scattering is a benefit, but it also causes difficult to investigate water structure changes owing to these weak signals. However, in these changes have been found an intensity decrease of the O–H stretching band (at 3250 cm^{-1}) relative to the C–H band (at 2938–2942 cm^{-1}) affected by water molecules and proteins interaction (Li-Chan and

Nakai 1991). RS could also be used to study food protein structure. For example, for secondary structure characterization, the –CO–NH– amide or peptide bond is the most commonly used, since it has several specificities and sensitivity of conformational vibrational modes at the amide I and III bands (Abbas *et al* 2012). Furthermore, RS can also be used for monitoring in food carbohydrate structure changes that were induced by processing or storage. RS has been used for the study of carbohydrate interactions with other food components, such as water (Li-Chan and Nakai 1991).

RS has been applied to analyze and quantify the lipid components in foods, including quantitative analysis of the unsaturation degree and the cis and trans isomers content, adulteration detection of various oils, polymorphism and chain packing characterization, and interactions monitoring of food components or changes caused by food processing or storage, e.g. isomerization or auto-oxidation. Dispersive laser RS and FT–RS were used as a rapid determination of unsaturation and cis and trans isomers content in food samples (Sadeghi-Jorabchi *et al* 1990, 1991, Yang *et al* 2005). Low-resolution RS has been developed to monitor olive oil oxidation status, where primary and secondary oxidation parameters (such as peroxide value) were measured using a rapid, non-destructive and direct method (Guzmán *et al* 2011).

Currently, RS is often coupled with Raman microspectroscopy, where they create an analytical tool suitable for the spatially resolved study of the food composition, including heterogeneous and food ingredients. The qualitative and quantitative analysis results can be developed using microspectroscopy. Based on their absorption unique pattern, various organic compounds and functional groups can be detected. Based on the absorption intensity, they could be applied for the determination of the relative concentration in the food sample. In addition, microscopic samples' size can be determined directly, at ambient pressure and temperature, in air, dry or wet, and in many various conditions without destroying the sample (Engelsen and Micklander 2003).

Confocal RS has been developed to give information regarding wheat grain composition and microstructure (Piot *et al* 2000). Here, the study has focused on the starchy endosperm protein composition and content and on the aleurone cell walls in ferulic acid and arabinoxylan derivatives. Moreover, it has been employed to monitor the protein content and structure evolution during grain development of various wheat varieties selected based on the hardness level and aptitude to the separation of peripheral layers during milling. RS is not only a powerful tool to identify cereal components, but it also provides information regarding these proteins' secondary structure and configuration. For example, the technique allows one to measure a non-specific conformation of wheat phospholipid transfer protein and to investigate the disulfide bridges' role in the α-helical structure stabilization. Furthermore, RS was employed to measure the puroindolines, and lipid-binding protein secondary structure and conformation in wheat (Le Bihan *et al* 1996).

2.7.2 Food authenticity and adulterations

In food authenticity and adulteration, RS coupled with chemometrics was used to determine honey adulterations by sugar and syrup solutions, and olive oil by vegetable oils or pomace olive oils, as well as meat and fish (Shurvell 2006). FT–RS coupled with multivariate analysis procedures were used to detect the adulteration level of virgin olive oil by some vegetable oils (e.g. corn, soybean, and raw olive residue oils) (Guzmán *et al* 2011). The dispersive RS coupled with multivariate data analysis was developed as a tool to determine adulteration of comminuted meat products with beef offal (liver, kidney, lung, and heart) (Zhao *et al* 2015). Raman spectral data for authenticated and adulterated beef burger samples prepared based on market formulations were tested by PLS-DA in order to identify authenticity and offal-adulterated beef burgers. PLS-DA results show classified authenticity accurately and adulteration with high accuracy. Moreover, PLS regression models were able to determine the total offal and added fat in the samples tested.

RS combined with chemometrics was employed as a tool to detect adulterants in honey, e.g. high fructose corn syrup (HFCS) and maltose syrup (MS) (Li *et al* 2012). Here, HFCS and MS were added in variable ratios to authentic honey samples. In order to remove the IR spectral background, the adaptive iteratively reweighted penalized least-squares (air-PLS) methodology was used for this purpose. PLS–LDA allowed for binary classification with 75.6%, 91.1% and 97.8% accuracy for authentic honey with HFCS- and MS-adulterated honey, HFCS-adulterated honey, and MS-adulterated honey, respectively.

2.8 NMR spectroscopy

NMR is one of the most reliable spectroscopic methods for 'high-throughput' and structural information on various molecular compounds of food products, such as fatty acids, amino acids, and sugars. It allows analyzing of complex compositional matrices in food products with high precision and accuracy. Moreover, any targetted metabolite amount in a mixture can be analyzed with less sample preparation.

NMR's main limitation in the past was its lack of sensitivity, but currently the continuous improvements in hardware and software have produced NMR with high sensitivity. Therefore, NMR allows a comprehensive metabolic profiles collection, which can be applied for food authentication and originality. Site-specific natural isotopic fractionation (SNIF-NMR), allows a robust natural molecule fingerprinting. A well-known SNIF-NMR application in food analysis is the development method by the EU in 1990 (e.g. EU regulations 2670/90, 2347/91) for the detection of a wine's geographical origin. An example is the profiling method based on non-targeted ^{1}H NMR analysis that was used for food geographical provenance determination (Longobardi *et al* 2013). Moreover, NMRS could be employed for determination of food adulteration, e.g. cane or corn sugar addition to maple syrup, adulteration of red wine with synthetic flavors sold as natural, anthocyanins, etc. The discrimination in the case of origin/adulteration in food based on NMR metabolic profile includes olive oils, coffee, honey, wines, spirits, vinegar, fish, and saffron.

NMR-based metabolomics has shown itself to be an effective and efficient tool for food authentication from raw materials to final product for quality control and for fraudulent labeling tracing (table 2.5). Moreover, NMR spectroscopy has proved by extensive study that it has benefits for food analysis, such as with olive oil, and is a useful tool for food quality analysis and authenticity. Generally, in NMR-based metabolomics, the two approaches have been developed, i.e. metabolic profiling and metabolic fingerprinting, where multivariate analysis has been applied for food authentication effectively (Dais and Hatzakis 2013).

2.8.1 ^{1}H NMR

NMRS has been developed for successful application to analyzing numerous food types for quality and authenticity, such as edible oils. For example, it has been used for sesame oil authenticity by using both fatty acid profiles and ^{1}H NMR spectra for the oils (Kim *et al* 2015). The NMR peaks based at σ^{13}C as an examined variable was signifying CH_3 groups in $n-3$ fatty acids, CH_2 groups between two C=C bonds, and protons from sesamin/sesamolin. The classification is based on effective variables by orthogonal projection to the latent variables in LDA. The results show that this authenticity approach of samples was accurately verified. Previously, this method has been proposed for detection of refined hazelnut oils in refined olive oils that allowed successful adulterant detection at low levels (10%). Therefore, the NMR spectrum can obtain an informative 'fingerprint' for various food products, as it provides crucial data regarding food variety and origin for classification analysis (Fragaki *et al* 2005).

^{1}H NMR fingerprinting was used to specify olive oils from various Mediterranean areas of their geographic origin and production year (Rezzi *et al* 2005). By applying PLS-DA to the PCA scores of ^{1}H NMR fingerprinting data, the results showed that olive oil samples can be discriminated based on the geographic origin and production year. ^{1}H NMR fingerprinting has been proposed for unsaponifiable matter in virgin olive oils for their characterization of geographical origin (Alonso-Salces *et al* 2010). LDA, PLS-DA, and SIMCA have been applied for NMR fingerprint multivariate analysis for olive oil samples from Italy, Spain, Greece, Turkey, Tunisia, and Syria. Then, they were evaluated based on their ^{1}H NMR spectra (the bulk oil, its unsaponifiable fraction, hydrocarbon, sterol, alcohol, and tocopherol sub-fractions of the unsaponifiable fraction), as chemical markers for the geographical characterization of oils. This study proved that ^{1}H NMR data for the bulk oil and its unsaponifiable fraction contained complementary data for geographical characterization analysis.

^{1}H NMR metabolomic fingerprinting combined with PCA and LDA has been used to determine Chinese rice (Huo *et al* 2017). PCA analysis allowed separate accurate rice samples from nine different Chinese provinces, while LDA afforded the specific variables for classification, such as sucrose, fructose, glucose, succinate polyphenols, asparagine, and trigonelline to be detected. The proposed method allowed discrimination among samples from different areas including temperature, latitude, or rainfall.

Table 2.5. Recent application of NMRS and chemometrics for food analysis.

Food sample	NMRS	Food analysis objective	References
Olive oil	H NMR	Olive oils classification	Rezzi *et al* (2005)
Olive oil	H NMR	Adulteration detection of refined olive oil with refined hazelnut oil	
Olive oils	C NMR	Identifying olive oils geographic origin	
Olive oils	H NMR C NMR	Olive oil characterization based on geographical, ecological and agronomic origin	
Virgin olive oil	H NMR C NMR	Virgin olive oil adulteration detection	
Virgin olive oil	H NMR C NMR	The unsaponifiable fraction of virgin olive oils fingerprinting for authentication	Alonso-Salces *et al* (2010)
Virgin olive oil	H NMR C NMR	Virgin olive oil quality assessment and authentication	Dais and Hatzakis (2013)
Extra virgin olive oil	P NMR	Adulteration detection in extra virgin olive oil	Fragaki *et al* (2005)
Edible Oils	P NMR H NMR	Edible oils classification and adulteration detection of virgin olive oils with seed oils	Vigli *et al* (2003)
Sesame oil	H NMR	Sesame oil authenticity verification	Kim *et al* (2015)
Cod liver oil	C NMR	Liver oil discrimination from wild/farmed cod and the geographic origin of cod identification	
Animal origin of heparin	H NMR C NMR	Animal origin authentication of heparin and low-molecular-weightHeparin (e.g. bovine and porcine species)	
Honey	H NMR	Honey authenticity	
Rice	H NMR	Rice discrimination from different geographical regions of China	Huo *et al* (2017)
Rice	H NMR	Rice type determination	
Wine	H NMR	NMR fingerprinting of wine	
Wine	H NMR C NMR	Wine analysis and authentication	
Wine	H NMR	Wine analysis: the grape variety, geographical origin, year of vintage	Godelmann *et al* (2013)
Wine	H NMR C NMR	Wine blends profiling	Imparato *et al* (2011)
Wine	H NMR C NMR	Authenticity determination, regional origin, and vintage of Slovenian wines	Ogrinc *et al* (2001)
Red wines	H NMR C NMR	Red wines classification	
Chinese wine	H NMR	Classification of Chinese wine varieties	
Chinese wine	H NMR C NMR	Chinese wines identification of the geographical origin	Jiang *et al* (2015)

Pistachios	H NMR C NMR	Pistachios authenticity control	
Green coffee	C NMR	Green coffee metabolomics classification	
Instant coffee	H NMR	Quality control and authentication of instant coffee	Charlton *et al* (2002)
Apple juice	H NMR	Apple varieties discrimination	Belton *et al* (1998)
Orange juice and pulp wash	H NMR	Marker compounds identification for discrimination between orange juice and pulp	Le Gall *et al* (2001)
Tomatoes	H NMR	Discriminating between conventionally and organically grown tomatoes	Hohmann *et al* (2014)
Grape	H NMR	Identification of grape variety, geographical origin, year of vintage	Godelmann *et al* (2013)

^{1}H NMR coupled with PCA and LDA methods have been used to identify German wines based on the grape variety, geographic origin and vintage year (Godelmann *et al* 2013). The method has enabled discriminating grape varieties from five wine-growing areas in southern and southwestern Germany. Moreover, it identified the geographic origin and vintage of wines from the regions of the largest German wine-products (Rheinpfalz, Mosel, Rheinhessen, Wurttemberg, and Baden) with 90% average accuracy.

NMR fingerprinting has been employed to classify monocultivar binary wine mixtures (Imparato *et al* 2011). LDA is used for classification, while ANN is used for quantification of the two monovarietal wines in each mixture of the irrelative proportions. The selected model could predict such proportions with 10% precision. SNIF-NMR combined with isotope ratio mass spectrometry (IRMS) methods has been applied for Slovenian wine authenticity and geographic origin (Ogrinc *et al* 2001). Based on PCA and LDA methods in combination with deuterium/proton isotopic ratios for the ethanol molecule methylene site, they allowed wine samples' discrimination between coastal and continental regions, where they are affected by weather events during grape ripening and harvesting. In addition, SNIF-NMRS and IRMS have been developed to identify the geographic origin of Chinese wines (Jiang *et al* 2015). LDA was applied for wine ethanol based on spectrum data from SNIF-NMRS, while the ratio for wine ethanol and the ratio for wine water were based on isotope data from IRMS, and the method enabled them to determine the regional wine origins from five different grape varieties of north Xinjiang.

NMR spectral data combined with PCA followed by LDA has been applied to classify three different manufacturers' products of instant spray dried coffees (Charlton *et al* 2002). Here the primary chemical marker (5-(hydroxymethyl)-2-furaldehyde) was used for differentiation and classification of coffees with excellent accuracy (100%) into three groups. ^{1}H NMR coupled with chemometrics has been applied for organically produced tomatoes' (*Solanum lycopersicum*) authentication (Hohmann *et al* 2014). The tomato samples of two different cultivars were analyzed from four different producers over a period of seven months. PCA enabled classification between organically and conventionally produced tomatoes, while LDA showed significant differences between growing regimes, and the classification of tomato samples by external validation confirmed 100% accuracy. Moreover, ^{1}H NMR spectral data coupled with PCA and LDA was applied for various apple juice varieties (Belton *et al* 1998). The covariances and a PCA correlation matrix have been used for different portions of intensity of the spectrum range. LDA enabled classification successfully, while PCA loadings plots allowed differentiation of the malic acid and sucrose based on two critical chemical level variables.

Last but not least, NMRS and LDA have been developed for detection of orange juice adulteration with pulp wash (Le Gall *et al* 2001). PCA and LDA have been applied to ^{1}H NMR spectra of juices (orange and pulp wash). The results showed the LDA model enabled classification of the training and validation sets to 94%. The PCA loading plots found that dimethylpropane plays a key role in the classification of the two juice types, especially with high pulp wash levels.

2.8.2 ^{13}C NMR

Similar to ^{1}H NMR, ^{13}C NMR has been successfully applied as a useful tool in food analysis. For example, by using high-gradient diffusion NMRspectroscopy, ^{13}C-NMR has been used to determine the diffusion coefficients (*D*) for extra virgin olive, seed and nut oils to determine the olive oils' adulterants (Šmejkalová and Piccolo 2010). The LDA has allowed detecting the LOD of adulteration levels, for soybean and sunflower oils these were 10%, while for hazelnut and peanut oils they were 30%. The LDA model was validated with randomly adulterated EVOO samples from an independent set with classification accuracy at 100%. The high accuracy of classification, along with the fast analysis, proved the high power potential of gradient diffusion NMR spectroscopy for rapid screening of olive oils adulteration.

2.8.3 ^{31}P NMR

Like ^{1}H NMR, ^{31}P NMR has been successfully developed for application to analyzing numerous food products for quality and authenticity. ^{31}P NMR spectroscopy has also been applied for analyses of three grades of olive oil samples from different olive varieties planted in Greece from various regions (Fragaki *et al* 2005). LDA was applied to five variables that allowed accurate classification of them. Here, LDA has classified successfully the three olive oil groups based on their quality. Discriminant analysis of different artificial mixtures was able to determine the extra virgin olive oil adulteration by refined or lampante olive oil with LOD at 5% (w/w).

The ^{31}P NMR spectra combined with ^{1}H NMR and LDA has been used to differentiate various vegetable oils, such as virgin olive, soybean, corn, sunflower, sesame, palm, rapeseed, walnut, hazelnut, almond, groundnut, coconut and safflower oils from various regions in Greece (Vigli *et al* 2003). 1,2-diglycerides, 1,3-diglycerides, fatty acid, iodine value, and acidity composition analysis based on ^{31}P

NMR and ^{1}H NMR spectral data were used as variables to build a mathematical model of LDA classification that was able to detect adulteration levels at LOD 5% (w/w).

2.9 Conclusions

The molecular spectroscopic methods seem to be a versatile tool in food quality analysis. They are applied to the analysis of food products qualitatively and quantitatively. Hence, the attractive features of these methods are their rapidity, simplicity, and ability to be applied to food analysis directly in various matrices, with rapid, minimal or even without any sample preparation. Development of novel, low-cost and portable, miniature or handheld spectroscopy instruments make them applicable in many fields, such as production lines, storage areas, outdoors, etc. Their potential dual measurement mode, transmittance, and reflectance make them useful in the various analytical tasks resolution. Furthermore, they give information on structure–functionality relationships, such as protein secondary structure, etc. These spectroscopy techniques can be employed individually or coupled with other analytical techniques as a detector such as in chromatography methods. IRS and RS are categorized as a rapid and simple tool in food analysis as no sample preparation is required, which in turn can be additionally applied as on-line analysis.

The spectroscopic techniques coupled with chemometric methods has created a tool for food analysis and characteristics, including authentication and adulteration of food. These methods for identifying food adulteration are mostly based on chemometric methods. Even though they are still in progress, some of these methods have already been integrated into benchtop and portable instruments. Further development of spectroscopy coupled with chemometrics as a tool in food analysis will be needed that is more robust and less susceptible to the matrix effect of the food sample.

The described methods in this chapter are mainly used to determine food compounds, (food components, ingredient, elements, and contaminants) and authenticity (adulteration type and level) of food samples, such as edible oils, etc. They can also be applied to classify cereal classes (wheat classes), measure flour quality, and the level of insects and fungal damage in food products, etc. However, firstly, the spectroscopic methods are used to measure the major food components content (carbohydrates, lipids, proteins and water), such that it is applied commonly in food analysis with excellent accuracy and sensitivity.

The next issue to be focussed on is the combination of spectroscopy with other analytical tools, as a tool for thermal, electrical or electrochemical measurement of food properties, such as the electronic nose, e-tongue, etc. This combination could increase the results reliability, and identification and classification accuracy without loss of simplicity, rapidity, or other advantages in food authenticity. For example, any additional data on the characteristics of the food, such as sample types, etc, will improve the accuracy of results to determine whether the food is authentic or fraud. In addition, the combination of molecular spectroscopic methods, with one or more of the physicochemical techniques, could obtain a significant improvement in identifying food authenticity and adulteration in the future. Food analysis in the future will require a rapid, simple and reliable method for revealing composition and ingredients of food products that is fundamental for safe and healthy food production to feed human needs.

Acknowledgments

This work was supported by Ditlitabmas, The Ministry of Research, Technology and Higher Education, the Republic of Indonesia through the World Class Research Grant 2019.

References

Abbas O and Baeten V 2013 Near-infra red spectroscopy *Spectroscopy Method in Food Analysis* ed A S Franca and L M L Nollet (Boca Raton, FL: CRC Press)

Abbas O, Dardenne P and Baeten V 2012 Near-infrared, mid-infrared, and raman spectroscopy *Chemical Analysis of Food: Techniques and Applications* (Amsterdam: Elsevier), pp 59–89

Al-Jowder O *et al* 1999 *Mid-Infrared Spectroscopy and Chemometrics for the Authentication of Meat Products* (Washington, DC: American Chemical Society)

Alamprese C *et al* 2013 Detection of minced beef adulteration with turkey meat by UV–vis, NIR and MIR spectroscopy *LWT - Food Sci. Technol.* **53** 225–32

Alamprese C and Casiraghi E 2015 Application of FT-NIR and FT-IR spectroscopy to fish fillet authentication *LWT - Food Sci. Technol.* **63** 720–5

Alonso-Salces R M *et al* 2010 Multivariate analysis of NMR fingerprint of the unsaponifiable fraction of virgin olive oils for authentication purposes *Food Chem.* **118** 956–65

Azcarate S M *et al* 2015 Modeling excitation-emission fluorescence matrices with pattern recognition algorithms for classification of Argentine white wines according grape variety *Food Chem.* **184** 214–9

Baker R S and Inventado P S 2014 Educational data mining and learning analytics *Learning Analytics* (New York: Springer), pp 61–75

Barbosa M F *et al* 2015 Identification of adulteration in ground roasted coffees using UV–Vis spectroscopy and SPA-LDA *LWT - Food Sci. Technol.* **63** 1037–41

Belton P S *et al* 1998 Application of chemometrics to the (1)H-NMR spectra of apple juices: discrimination between apple varieties *Food Chem.* **61** 207–13

Berrueta L A, Alonso-Salces R M and Héberger K 2007 Supervised pattern recognition in food analysis *J. Chromatogr.* A **1158** 196–214

Bertran E *et al* 1999 Determination of olive oil free fatty acid by fourier transform infrared spectroscopy *J. Am. Oil Chem. Soc.* **76** 611–6

Biji K B *et al* 2015 Smart packaging systems for food applications: a review *J. Food Sci. Technol.* **52** 6125–35

Blanco M and Villarroya I 2002 NIR spectroscopy: a rapid-response analytical tool *Trends Anal. Chem.* **21** 240–50

Callao M P and Ruisánchez I 2018 An overview of multivariate qualitative methods for food fraud detection *Food Control* **86** 283–93

Charlton A J, Farrington W H H and Brereton P 2002 *Application of ^1H NMR and Multivariate Statistics for Screening Complex Mixtures: Quality Control and Authenticity of Instant Coffee* (Washington, DC: American Chemical Society)

Che Man Y B, Moh M H and van de Voort F R 1999 Determination of free fatty acids in crude palm oil and refined-bleached-deodorized palm olein using Fourier transform infrared spectroscopy *J. Am. Oil Chem. Soc.* **76** 485–90

Corneli S and Maragos C M 1998 Capillary electrophoresis with laser-induced fluorescence: method for the mycotoxin ochratoxin A *J. Agric. Food Chem.* **46** 3161–65

Craig A P, Franca A S and Irudayaraj J 2013 Surface-enhanced Raman spectroscopy applied to food safety *Annu. Rev. Food Sci. Technol.* **4** 369–80

Dais P and Hatzakis E 2013 Quality assessment and authentication of virgin olive oil by NMR spectroscopy: a critical review *Anal. Chim. Acta* **765** 1–27

Dalle Zotte A *et al* 2014 Authentication of raw and cooked freeze-dried rainbow trout (Oncorhynchus mykiss) by means of near infrared spectroscopy and data fusion *Food Res. Int.* **60** 180–8

Danezis G P *et al* 2016 Food authentication: Techniques, trends and emerging approaches *Trends Anal. Chem.* **85** 123–32

Dankowska A, Domagała A and Kowalewski W 2017 Quantification of *Coffea arabica* and *Coffea canephora* var. robusta concentration in blends by means of synchronous fluorescence and UV–Vis spectroscopies *Talanta* **172** 215–20

Dankowska A, Małecka M and Kowalewski W 2014 Application of synchronous fluorescence spectroscopy with multivariate data analysis for determination of butter adulteration *Int. J. Food Sci. Technol.* **49** 2628–34

Dankowska A, Małecka M and Kowalewski W 2015 Detection of plant oil addition to cheese by synchronous fluorescence spectroscopy *Dairy Sci. Technol.* **95** 413–24

Di Anibal C V *et al* 2012 Surface Enhanced Raman Spectroscopy (SERS) and multivariate analysis as a screening tool for detecting Sudan I dye in culinary spices *Spectrochim. Acta Part A, Mol. Biomol. Spectrosc.* **87** 135–41

Ding X, Ni Y and Kokot S 2015 NIR spectroscopy and chemometrics for the discrimination of pure, powdered, purple sweet potatoes and their samples adulterated with the white sweet potato flour *Chemometr. Intell. Lab. Syst.* **144** 17–23

Diniz P H G D *et al* 2016 Using UV–Vis spectroscopy for simultaneous geographical and varietal classification of tea infusions simulating a home-made tea cup *Food Chem.* **192** 374–79

Dobrucka R and Cierpiszewski R 2014 Active and intelligent packaging food research and development- a review *Pol. J. Food Nutr. Sci.* **64** 7–15

Downey G 1996 *J. Near Infrared Spectrosc.* **4** 47–61

Dubois J *et al* 1996 Quantitative Fourier transform infrared analysis for anisidine value and aldehydes in thermally stressed oils *J. Am. Oil Chem. Soc.* **73** 787–94

Durán Merás I *et al* 2018 Detection and quantification of extra virgin olive oil adulteration by means of autofluorescence excitation-emission profiles combined with multi-way classification *Talanta* **178** 751–62

El Darra N *et al* 2017 Food fraud detection in commercial pomegranate molasses syrups by UV–VIS spectroscopy, ATR-FTIR spectroscopy and HPLC methods *Food Control* **78** 132–7

Ellis D I *et al* 2015 Point-and-shoot: rapid quantitative detection methods for on-site food fraud analysis—moving out of the laboratory and into the food supply chain *Anal. Methods* **7** 9401–14

Engelsen S B 1997 Explorative spectrometric evaluations of frying oil deterioration *J. Am. Oil Chem. Soc.* **74** 1495

Engelsen S B and Micklander E 2003 Vibrational microspectroscopy of food. Raman vs. FT-IR *Trends Food Sci. Technol.* **14** 50–7

Esteki M, Shahsavari Z and Simal-Gandara J 2018 Use of spectroscopic methods in combination with linear discriminant analysis for authentication of food products *Food Control* **91** 100–12

Fashi A, Yaftian M R and Zamani A 2017 Electromembrane extraction-preconcentration followed by microvolume UV–Vis spectrophotometric determination of mercury in water and fish samples *Food Chem.* **221** 714–20

Fragaki G *et al* 2005 *Detection of Extra Virgin Olive Oil Adulteration with Lampante Olive Oil and Refined Olive Oil Using Nuclear Magnetic Resonance Spectroscopy and Multivariate Statistical Analysis* (Washington, DC: American Chemical Society)

Gall G L, Puaud M and Colquhoun I J 2001 *Discrimination between Orange Juice and Pulp Wash by ^1H Nuclear Magnetic Resonance Spectroscopy: Identification of Marker Compounds* (Washington, DC: American Chemical Society)

Georgouli K, Martinez Del Rincon J and Koidis A 2017 Continuous statistical modelling for rapid detection of adulteration of extra virgin olive oil using mid infrared and Raman spectroscopic data *Food Chem.* **217** 735–42

Godelmann R *et al* 2013 Targeted and nontargeted wine analysis by ^{1}H NMR spectroscopy combined with multivariate statistical analysis. differentiation of important parameters: grape variety, geographical origin, year of vintage *J. Agric. Food Chem.* **61** 5610–9

Gonçalves R P, Março P H and Valderrama P 2014 Thermal edible oil evaluation by UV–Vis spectroscopy and chemometrics *Food Chem.* **163** 83–6

Guimet F, Ferré J and Boqué R 2005 Rapid detection of olive–pomace oil adulteration in extra virgin olive oils from the protected denomination of origin 'Siurana' using excitation–emission fluorescence spectroscopy and three-way methods of analysis *Anal. Chim. Acta* **544** 143–52

Gurdeniz G and Ozen B 2009 Detection of adulteration of extra-virgin olive oil by chemometric analysis of mid-infrared spectral data *Food Chem.* **116** 519–25

Gurdeniz G, Tokatli F and Ozen B 2007 Differentiation of mixtures of monovarietal olive oils by mid-infrared spectroscopy and chemometrics *Eur. J. Lipid Sci. Technol.* **109** 1194–202

Guzmán E *et al* 2011 Application of low-resolution Raman spectroscopy for the analysis of oxidized olive oil *Food Control* **22** 2036–40

Harwood J and Aparicio R 2000 *Handbook of Olive Oil: Analysis and Properties* (Berlin: Springer)

Herrero A M 2008 Raman spectroscopy a promising technique for quality assessment of meat and fish: A review *Food Chem.* **107** 1642–51

Hohmann M *et al* 2014 ^{1}H NMR profiling as an approach to differentiate conventionally and organically grown tomatoes *J. Agric. Food Chem.* **62** 8530–40

Huo Y *et al* 2017 ^{1}H NMR-based metabolomics for discrimination of rice from different geographical origins of China *J. Cereal Sci.* **76** 243–52

Hussain M N *et al* 2014 Feasibility study of detecting canola oil adulteration with palm oil using NIR spectroscopy and multivariate analysis *2014 IEEE Proc. of Int. Conf. on Information, Communication Technology and System (ICTS)* pp 111–4

Imparato G *et al* 2011 Nuclear magnetic resonance profiling of wine blends *J. Agric. Food Chem.* **59** 4429–34

Jiang W *et al* 2015 The application of SNIF-NMR and IRMS combined with C, H and O isotopes for detecting the geographical origin of Chinese wines *Int. J. Food Sci. Technol.* **50** 774–81

Johnson R 2014 Food fraud and 'Economically motivated adulteration' of food and food ingredients *Congressional Research Service* 1–56

Karoui R and Blecker C 2011 Fluorescence spectroscopy measurement for quality assessment of food systems—a review *Food Bioprocess Technol.* **4** 364–86

Kim J *et al* 2015 Combined analysis of stable isotope, ^{1}H NMR, and fatty acid to verify sesame oil authenticity *J. Agric. Food Chem.* **63** 8955–65

Kuswandi B *et al* 2011 Smart packaging: Sensors for monitoring of food quality and safety *Sens. Instrum. Food Qual. Saf.* **5** 137–46

Kuswandi B *et al* 2015 Pork adulteration in commercial meatballs determined by chemometric analysis of NIR Spectra *J. Food Meas. Charact.* **9** 313–23

Kuswandi B 2016 Nanotechnology in food packaging *Nanoscience in Food and Agriculture 1* ed S Ranjan, D Nandita and E Lichtfouse 1st edn (Berlin: Springer) p 151

Kyriakidis N B and Skarkali P 2000 Fluorescence spectra measurement of olive oil and other vegetable oils *J. AOAC Int.* **83** 1435–39

Lakowicz J R E 1999 *Principles of Fluorescence Spectroscopy* (New York: Springer)

Larkin P 2011 Environmental dependence of vibrational spectra *Infrared and Raman Spectroscopy* (Amsterdam: Elsevier) pp 55–62

Lazzerini C and Domenici V 2017 Pigments in extra-virgin olive oils produced in Tuscany (Italy) in different years *Foods* **6**

Le Bihan T *et al* 1996 Determination of the secondary structure and conformation of puroindolines by infrared and raman spectroscopy *Biochemistry* **35** 12712–22

Li-Chan E C Y *et al* 2006 Vibrational spectroscopy of food and food products *Handbook of Vibrational Spectroscopy* ed P R Griffiths (Chichester: Wiley)

Li-Chan E and Nakai S 1991 Raman spectroscopic study of thermally and/or dithiothreitol induced gelation of lysozyme *J. Agric. Food Chem.* **39** 1238–45

Lin C-A, Ayvaz H and Rodriguez-Saona L E 2014 Application of portable and handheld infrared spectrometers for determination of sucrose levels in infant cereals *Food Anal. Methods* **7** 1407–14

Li P, Zhang Q and Zhang W 2009 Immunoassays for aflatoxins *Trends Anal. Chem.* **28** 1115–26

Li S *et al* 2012 Detection of honey adulteration by high fructose corn syrup and maltose syrup using Raman spectroscopy *J. Food Compos. Anal.* **28** 69–74

Lohumi S *et al* 2015 A review of vibrational spectroscopic techniques for the detection of food authenticity and adulteration *Trends Food Sci. Technol.* **46** 85–98

Longobardi F *et al* 2013 Non-targeted ^{1}H NMR fingerprinting and multivariate statistical analyses for the characterisation of the geographical origin of Italian sweet cherries *Food Chem.* **141** 3028–33

Mahesh S *et al* 2008 Feasibility of near-infrared hyperspectral imaging to differentiate Canadian wheat classes *Biosyst. Eng.* **101** 50–7

Maragos C M 1997 Detection of the mycotoxin fumonisin B_1 by a combination of immunofluorescence and capillary electrophoresis *Food Agric. Immunol.* **9** 147–57

Maragos C M *et al* 2001 Fluorescence polarization as a means for determination of fumonisins in maize *J. Agric. Food Chem.* **49** 596–602

Maragos C M and Plattner R D 2002 Rapid fluorescence polarization immunoassay for the mycotoxin deoxynivalenol in wheat *J. Agric. Food. Chem.* **50** 1827–32

Maria Porcu O 2018 The importance of UV–Vis spectroscopy: application in food products characterization *Scholarly J. Food Nutr.* **1** 59–62

Martins A R *et al* 2017 Discrimination of whisky brands and counterfeit identification by UV–Vis spectroscopy and multivariate data analysis *Food Chem.* **229** 142–51

Mejri M *et al* 2005 Effects of some additives on wheat gluten solubility: A structural approach *Food Chem.* **92** 7–15

Milanez K D T M *et al* 2017 Multivariate modeling for detecting adulteration of extra virgin olive oil with soybean oil using fluorescence and UV–Vis spectroscopies: A preliminary approach *LWT - Food Sci. Technol.* **85** 9–15

Miller J and Miller J 2010 *Statistics and Chemometrics for Analytical Chemistry* 6th edn (Harlow: Pearson)

Moh M H *et al* 1999 Quantitative analysis of palm carotene using Fourier transform infrared and near infrared spectroscopy *J. Am. Oil Chem. Soc.* **76** 249

Moros J *et al* 2009 Preliminary studies about thermal degradation of edible oils through attenuated total reflectance mid-infrared spectrometry *Food Chem.* **114** 1529–36

Morsy N and Sun D-W 2013 Robust linear and non-linear models of NIR spectroscopy for detection and quantification of adulterants in fresh and frozen-thawed minced beef *Meat Sci.* **93** 292–302

Nakai S and Horimoto Y 2000 Fluorescence spectroscopy in food analysis *Encyclopedia of Analytical Chemistry* (Chichester: Wiley)

Navarra G *et al* 2011 Thermal oxidative process in extra-virgin olive oils studied by FTIR, rheology and time-resolved luminescence *Food Chem.* **126** 1226–31

Nzai J M and Proctor A 1998 Determination of phospholipids in vegetable oil by fourier transform infrared spectroscopy *J. Am. Oil Chem. Soc.* **75** 1281–9

Ogrinc N *et al* 2001 *Determination of Authenticy, Regional Origin, and Vintage of Slovenian Wines Using a Combination of IRMS and SNIF-NMR Analyses* (Washington, DC: American Chemical Society)

Oppermann U 2004 *Spectroscopy in Food Analysis G.I.T. Laboratory J.* **4** 2–4

Osborne B G 2000 Near-infrared spectroscopy in food analysis *Encyclopedia of Analytical Chemistry* (Chichester: Wiley) pp 1–14

Otles S and Ozyurt V H 2015 Instrumental food analysis *Handbook of Food Chemistry* (Berlin: Springer) pp 165–87

Palacios-Morillo A *et al* 2013 Differentiation of tea varieties using UV–Vis spectra and pattern recognition techniques *Spectrochim. Acta, Part A* **103** 79–83

Paradkar M, Sivakesava S and Irudayaraj J 2003 Discrimination and classification of adulterants in maple syrup with the use of infrared spectroscopic techniques *J. Sci. Food Agric.* **83** 714–21

Pena-Pereira F *et al* 2011 Advances in miniaturized UV–Vis spectrometric systems *Trends Anal. Chem.* **30** 1637–48

Piot O, Autran J-C and Manfait M 2000 Spatial distribution of protein and phenolic constituents in wheat grain as probed by confocal Raman microspectroscopy *J. Cereal Sci.* **32** 57–71

Ram M S, Seitz L M and Dowell F E 2004 Natural fluorescence of red and white wheat kernels *Cereal Chem. J.* **81** 244–8

Reis N *et al* 2017 Simultaneous detection of multiple adulterants in ground roasted coffee by ATR-FTIR spectroscopy and data fusion *Food Anal. Methods* **10** 2700–9

Rezzi S *et al* 2005 Classification of olive oils using high throughput flow ^{1}H NMR fingerprinting with principal component analysis, linear discriminant analysis and probabilistic neural networks *Anal. Chim. Acta* **552** 13–24

Rodriguez-Saona L and Allendorf M 2011 Use of FTIR for rapid authentication and detection of adulteration of food *Annu. Rev. Food Sci. Technol.* **2** 467–83

Rohman A and Man Y B C 2010 Fourier transform infrared (FTIR) spectroscopy for analysis of extra virgin olive oil adulterated with palm oil *Food Res. Int.* **43** 886–92

Rohman A and Man Y B C 2011 The use of Fourier transform mid infrared (FT-MIR) spectroscopy for detection and quantification of adulteration in virgin coconut oil *Food Chem.* **129** 583–8

Ruoff K *et al* 2005 *Authentication of the Botanical Origin of Honey by Front-Face Fluorescence Spectroscopy. A Preliminary Study* (Washington, DC: American Chemical Society)

Sadeghi-Jorabchi H *et al* 1990 Determination of the total unsaturation in oils and margarines by fourier transform raman spectroscopy *J. Am. Oil Chem. Soc.* **67** 483–6

Sadeghi-Jorabchi H *et al* 1991 Quantitative analysis of oils and fats by Fourier transform Raman spectroscopy *Spectrochim. Acta Part A: Mol. Spectrosc.* **47** 1449–58

Santos P M, Pereira-Filho E R and Rodriguez-Saona L E 2013 Rapid detection and quantification of milk adulteration using infrared microspectroscopy and chemometrics analysis *Food Chem.* **138** 19–24

Sayago A, Morales M T and Aparicio R 2004 Detection of hazelnut oil in virgin olive oil by a spectrofluorimetric method *Eur. Food Res. Technol.* **218** 480–3

Shurvell H F 2006 Spectra-structure correlations in the mid- and far-infrared *Handbook of Vibrational Spectroscopy* ed J M Chalmers (Chichester: Wiley)

Sikorska E *et al* 2005 Classification of edible oils using synchronous scanning fluorescence spectroscopy *Food Chem.* **89** 217–25

Singh C B *et al* 2009 Detection of insect-damaged wheat kernels using near-infrared hyperspectral imaging *J. Stored Prod. Res.* **45** 151–8

Singh C B *et al* 2010 Identification of insect-damaged wheat kernels using short-wave near-infrared hyperspectral and digital colour imaging *Comput. Electron. Agric.* **73** 118–25

Skoog D A, Holler F J and Nieman T A 1998 *Principles of Instrumental Analysis* (Philadelphia, PA: Saunders)

Šmejkalová D and Piccolo A 2010 High-power gradient diffusion NMR spectroscopy for the rapid assessment of extra-virgin olive oil adulteration *Food Chem.* **118** 153–8

Stuart B 2004 *Infrared Spectroscopy: Fundamental and Applications* (Philadelphia, PA: Saunders)

Suhandy D and Yulia M 2017 Peaberry coffee discrimination using UV–visible spectroscopy combined with SIMCA and PLS-DA *Int. J. Food Prop.* **20** S331–9

Symons S J and Dexter J E 1991 Computer analysis of fluorescence for the measurement of flour refinement as determined by flour ash content flour grade color and tristimulus color measurements *Cereal Chem.* **68** 454–60

Tan J *et al* 2017 Synchronous front-face fluorescence spectroscopy for authentication of the adulteration of edible vegetable oil with refined used frying oil *Food Chem.* **217** 274–80

Tan J, Li R and Jiang Z-T 2015 Chemometric classification of Chinese lager beers according to manufacturer based on data fusion of fluorescence, UV and visible spectroscopies *Food Chem.* **184** 30–6

van de Voort F R *et al* 1992 A rapid FTIR quality control method for fat and moisture determination in butter *Food Res. Int.* **25** 193–8

van de Voort F R *et al* 1994 The determination of peroxide value by Fourier transform infrared spectroscopy *J. Am. Oil Chem. Soc.* **71** 921–6 van de Voort F R, Sedman J and Ismail A A 1993 A rapid FTIR quality-control method for determining fat and moisture in high-fat products *Food Chem.* **48** 213–21

Vasconcelos M *et al* 2015 Study of adulteration of extra virgin olive oil with peanut oil using FTIR spectroscopy and chemometrics *Cogent Food Agric.* **1** 1–13

Vigli G *et al* 2003 *Classification of Edible Oils by Employing ^{31}P and ^{1}H NMR Spectroscopy in Combination with Multivariate Statistical Analysis. A Proposal for the Detection of Seed Oil Adulteration in Virgin Olive Oils* (Washington, DC: American Chemical Society)

Yang D and Ying Y 2011 Applications of Raman spectroscopy in agricultural products and food analysis: a review *Appl. Spectrosc. Rev.* **46** 539–60

Yang H, Irudayaraj J and Paradkar M 2005 Discriminant analysis of edible oils and fats by FTIR, FT-NIR and FT-Raman spectroscopy *Food Chem.* **93** 25–32

Zandomeneghi M 1999 *Fluorescence of Cereal Flours* (Washington, DC: American Chemical Society)

Zhao M, Downey G and O'Donnell C P 2015 Dispersive Raman spectroscopy and multivariate data analysis to detect offal adulteration of thawed beefburgers *J. Agric. Food Chem.* **63** 1433–41

Chapter 3

Nuclear magnetic resonance spectroscopy applications in the food industry

Y Parlak

Nuclear magnetic resonance spectroscopy (NMR) is one of the most powerful techniques for determining the structure of food. High resolution NMR, low field NMR, NMR—Mobile Universal Surface Explorer and 2D NMR techniques are all utilized successfully in the food industry. NMR is a useful technique for determination of food quality. In addition, NMR is of benefit for determination of geographical origin of foods, oxidative stability of oil varieties and metabolite determination. This review includes the specific NMR applications in the analysis of foods such as food emulsions, the beverage industry, milk and dairy products, meat research, carbohydrate analyses, and fruit and vegetable research.

3.1 Nuclear magnetic resonance spectroscopy

NMR is a useful technique for determination of the structure of molecules. It defines the carbon–hydrogen framework of an organic compound. NMR refers to the behaviour of atoms subjected to a magnetic field. The phenomenon was first described in 1946 by Bloch and Purcell. Atoms with an odd mass number such as ^{1}H, ^{31}P and ^{13}C possess the quantum property of 'spin' and behave as dipoles aligning along the axis of an applied magnetic field (Tognarelli *et al* 2015). NMR is based on the magnetic moments of some atomic nuclei (e.g. the nuclei of ^{1}H, ^{13}C, ^{15}N) which enable them to reorient when exposed to a fixed external magnetic field and absorb radiofrequency energy (Gil and Rodrigues 2008). NMR is particularly useful for identifying variation between the chemical compositions of solutions, and can be used as a fingerprinting method or to identify marker compounds (Charlton *et al* 2002).

The atomic nucleus is a spinning charged particle, and it generates a magnetic field. Without an external applied magnetic field, the nuclear spins are random and spin in random directions. But, when an external magnetic field is present, the nuclei

doi:10.1088/2053-2563/ab4428ch3

align themselves either with or against the field of the external magnet (Proton Nuclear Magnetic Resonance Spectroscopy (H-NMR) 2019). In NMR spectroscopy, the absorption bands are referred to as 'peaks', and the graph obtained by marking the frequencies against peaks formed as a result of absorption is called 'NMR spectra'.

An NMR spectrophotometer consists of four essential components.

1. A powerful magnet (provides a magnetic field into which the sample is placed).
2. One or more radio frequency transmitters.
3. A radio frequency receiver.
4. A recorder.

An NMR spectrum gives the following information:

(a) The number of peaks indicate different types of nucleus.
(b) The location of the peak indicates the type of nucleus and chemical environment.
(c) The relative areas of the peaks give the relative number of each type of nucleus.
(d) Disruption in the peak distinguishes those affected nuclei from each other (Keeler 2010).

The most common types of NMR are proton and carbon-13 NMR spectroscopy, and these are used in food science successfully. The ^{13}C CP-MAS NMR method is preferred due to its ability to determine distribution of ^{13}C among major categories of organic molecules without the need for solvent extraction (Yekta *et al* 2019). ^{1}H nuclear magnetic resonance (NMR) spectroscopy gives an insight on the molecular level of a food similar to a 'fingerprint' within a single spectrum (Ackermann *et al* 2019). NMR has been used for quality and quantitative food analysis in many fields, for instance, the characterisation of the geographic origin of juice, geographic origin of milk and cheese, grape origin, and variety of wine, or the botanical origin of honey.

31-Phosphorus nuclear magnetic resonance spectroscopy (^{31}P NMR) has been used to determine the headgroup composition of phospholipids (PLs) in complex lipid mixtures. However, chemical shifts of PL headgroups in ^{31}P NMR spectra are distributed over a rather small range of only about 2 ppm. As a consequence, ^{31}P resonances from different PL head groups often overlap. Such problems can be solved by means of two-dimensional (2D) NMR approaches (Kaffarnik *et al* 2013).

3.2 Use of NMR techniques in food science

Foods are complex systems and NMR is a simple and fast method for illuminating their structures. NMR spectroscopy is used in several food systems for different purposes. Food emulsions, the beverage industry, dairy products, meat research, carbohydrate analyses, fruit and vegetable research are all examples.

^{1}H, ^{13}C and ^{31}P have been extensively used in food science. For the same sample, the researcher can choose among nuclei to detect different characteristics, or to improve the spectrum. For instance, milk fat lipids can be analysed from either ^{1}H or ^{13}C spectra, the latter giving better resolution through lower signal-to-noise (Belloque and Ramos 1999).

Some NMR techniques are utilized in the food industry, such as high resolution NMR, low field NMR, NMR—Mobile Universal Surface Explorer, solid state NMR, and magic-angle spinning (MAS) NMR.

In order to obtain the NMR spectrum of a substance, it is necessary to prepare the solution by dissolving a sample in a suitable solvent. Therefore, signals will be observed for the solvent and this must be accounted for in solving spectral problems. Generally, deuterium oxide (heavy water, water-d_2, D_2O) has been used as solvent for the dissolution of internal standard and food sample during quantification experiments by NMR. But deuterated water has been used with three-methyl sodium propionate (TSP) in some foods.

3.2.1 Application of NMR spectroscopy to milk and dairy products

NMR spectroscopy technique is useful for qualitative and quantitative analysis, monitoring reactions *in vivo*, isotopic analysis, study of the physical state of milk fat and water, and structural characterization of proteins. NMR techniques have been used for characterization of the ripening and geographical characterisation of cheeses such as Parmigiano Reggiano, Mozzarella, Grana Padano and Circassian cheese. The low resolution nuclear magnetic resonance technique has been used to identify the water fractions in cheeses. ^{1}H NMR spectra are very complex, and a strong overlap of the resonances occurs. The unequivocal assignment is obtained by 2D NMR COSY experiments where scalar interactions between adjacent groups of the same molecules are revealed (Curtıs *et al* 2000).

Brescia *et al* (2005) worked on characterization of the geographical origin of buffalo milk and mozzarella cheese by NMR. Only the coupling of the isotopic parameters with NMR data determined on the aqueous mozzarella extracts allowed good results concerning geographical origin discrimination.

Shintu and Caldarelli (2005) took advantage of high-resolution MAS NMR for the characterization of the ripening of Parmigiano Reggiano cheese. The high quality spectra enabled the identification and attribution of most of the water soluble component. In this research, standard TOCSY HSQC and HMBC spectra led to unambiguous assignment of 23 amino and organic acids spin systems while six other spin systems of unknown compounds were detected.

High resolution NMR experiments allow all the amino acids present in the cheese to be quantitatively and qualitatively evaluated. High resolution NMR allows quantitative determination of the free amino acids content (Curtıs *et al* 2000). Figure 3.1 shows the ^{1}H spectrum of free amino acids and organic acids in cheese obtained by ^{1}H NMR (Parlak 2016). Alanine, arginine, glycine, glutamine, isoleucine, leucine, lysine, phenylalanine, proline, serine, valine, lactic acid, citric acid and acetic acid were identified in Circassian cheese with support from NMR.

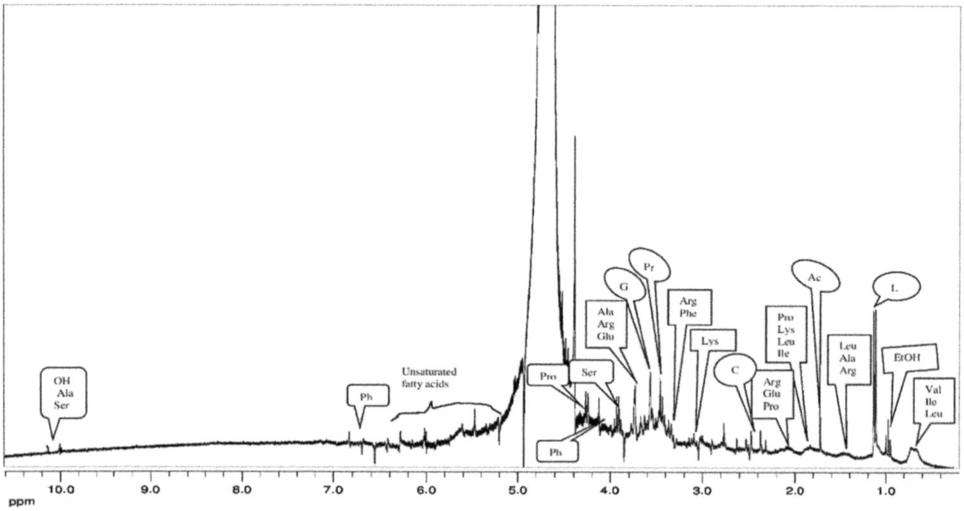

Figure 3.1. ^{1}H spectrum of free amino acids in Circassian cheese, at 3 months of ripening, obtained by ^{1}H NMR nuclear magnetic resonance (Parlak 2016).

An amino acid map of cheeses can be set by using NMR spectra. Table 3.1 identified NMR characteristics of cheese (Parlak 2016). Glutamic acid, lactate, asparagine, phenylalanine, glycine, glutamine, isoleucine, histidine, leucine, lysine, methionine, proline, serine, threonine and valine were determined in the study of the amino acid profile of Grana Padano cheese by Curtıs *et al* (2000). Studies have shown that determinable amounts of free amino acids occur on the 40th and 60th days of storage and some amino acids occur on the 240th day. The amino acid profile is an indicator which shows the significant variations in ripening period.

Furthermore, ^{1}H NMR method was developed to analyze histamine in cheeses. Schievano *et al* (2009) determined, for the first time, an accurate NMR determination of histamine down to very low concentrations in a very short time. The procedure is simple because the acid extract is analyzed directly, without any need for further filtration, derivatization, or other manipulation.

Kaffarnik *et al* (2013) profiled PLs in cheese and fish with one-dimensional (1D) ^{31}P NMR and two-dimensional (2D) ^{31}P, ^{1}H COSY NMR. Phosphatidylcholine, alkyl ether-linked phosphatidylcholine, lysophosphatidylcholine, lysophosphatidylcholine plasmalogen, phosphatidylethanolamine, alkyl ether-linked phosphatidylethanolamine, phosphatidylinositol, sphingomyelin, dihydrosphingomyelin, phosphatidic acid, phosphatidylglycerol and lysophosphatidic acid were observed at 1D ^{31}P NMR spectra in cheese and fish.

Maruyama *et al* (2014) have studied complementary analyses of fractal and dynamic water structures in protein–water mixtures and cheeses. According to this study, the diffusion coefficient obtained from NMR decreased with increasing protein concentration.

Table 3.1. Characteristics of ^{1}H NMR signals observable in Circassian cheese (Parlak 2016).

Compounds	Chemical shifts δ ^{1}H (ppm)	Multiplicity[a]	Assignment
Alanine (Ala)	1.45	d	β-CH_3
	3.75	q	α-CH
Arginine (Arg)	~1.60–1.80	m (overlap)	β, β'-CH_2
	3.29	t	δ-CH_2
	3.5	m	α-CH
Glycine (Gly)	3.54	s	α-CH
Glutamine (Glu)	2.2	d–t	β and γ-CH_2
	3.71	dd (t) broad	α-CH
Isoleucine (Ile)	~0.7	t (overlap)	δ-CH_3
	~1.05	m (broad)	γ-CH_3
	1.82	m (overlap)	β-CH
	3.65	m (overlap)	α-CH
Leucine (Leu)	~0.7–0.8	d (overlap)	δ, δ'-CH_3
	~1.5	m (broad)	γ-CH
	~1.8	m (overlap)	β-CH_2
	~3.7	m (overlap)	α-CH
Lysine (Lys)	~1.9	M	β-CH_2
	~3.05	T	ε-CH_2
	~3.75	T	α-CH
Phenylalanine (Phe)	~3.3		β-CH_2
	~4.0	dd	α-CH
	~6.7–6.9	dd(t)	ring-CH
Proline (Pro)	1.95	M	γ-CH_2
	~2.0	M	β-CH_2
	~3.4	t–m	δ-CH_2
	~4.2–4.3	dd(t)	α-CH
Serine (Ser)	~3.8	dd	α-CH
	~4.0	d	β-CH_2
Valine (Val)	~0.7–0.9	d	γ,γ'-CH_3
	~1.5	d	α-CH
	~1.8	Unresolved	β-CH
Lactic acid (Lac)	1.11	d	β-CH
	3.90	q	α-CH
Citric acid (Cit)	2.31	d	–CH_2
	2.47	d	–CH_2
Acetic acid (Ace)	1.70	s	–CH_3

[a] Multiplicity: s, singlet; d, doublet; t, triplet; q, quartet; dd, doublet of doublets; m, multiplet.

3.2.2 Application of NMR spectroscopy to the beverage industry

NMR spectroscopy is used for discrimination of fruit juice. Belton *et al* (1998) used ^{1}H NMR for discrimination between apple juices produced from different varieties (Spartan, Bramley, Russet), and under optimum conditions a 100% success rate has been achieved. The region between 2.5 and 5.5 ppm contains signals from the components of highest concentration: sucrose, fructose, glucose and malic acid. NMR has been utilized for discrimination of five different mango cultivars, Awin, Carabao, Keitt, Kent, and Nam Dok Mai, using metabolic analysis with band-selective excitation NMR spectra. In this research, a combination of unsupervised principal component analysis (PCA) with low-field region ^{1}H NMR spectra has provided a good discriminant model of the five mango cultivars. Various minor components have been identified in the mango juice by using F2-selective 2D NMR spectra (Koda *et al* 2012).

NMR is successfully being used to characterize wine and find an association of wine metabolite with environmental and fermentative factors in the vineyard and for making wine (Hong 2012). A number of metabolites having different chemical characteristics in grape juice and wine have been identified using ^{1}H NMR spectroscopy by amino acids (leucine, isoleucine, valine, threonine, alanine, arginine, glutamine, γ-aminobutyric acid, proline, tyrosine, and amino acid derivatives), organic acids (succinate, acetate, malate, tartarate, and citrate), sugars (α- and β-glucose, fructose, sucrose, and unknown carbohydrate), 2,3-butanediol, glycerol, 2-phenylethanol, trigonelline, and phenylpropanoids (cis/trans-caftaric acid, cis/trans-caffeoyl malate, and cis/trans-coutaric acid) (Hong 2011).

In beer analysis, it is possible to distinguish between major beer types and to detect the geographical origin of beer by NMR (Kuballa *et al* 2018). ^{1}H NMR spectroscopy combined with chemometrics was employed to discriminate lager beer samples from two different classes, according to their style and information provided on the label by Silva *et al* (2019). As there have no studies concerning the discrimination of beers of the same type that differ only in style, they have used ^{1}H NMR spectroscopy.

An NMR and chemometric analytical approach to classify beers according to their brand identity was developed by Mannina *et al* (2010). Beers have been analyzed by ^{1}H NMR spectroscopy and selected NMR signals have been used to build classification models. In addition, chemical composition of beers has been also obtained.

3.2.3 Application of NMR spectroscopy to the food oil industry

^{1}H and ^{13}C NMR is used for similarity and differential NMR spectroscopy in metabolomics of vegetable oils. It has been demonstrated that the calculation of similarity and differential NMR spectroscopy permit the classification of vegetable oils and has been used for the detection of adulteration by Schripsema (2019). Differential NMR spectroscopy can be used to detect marker compounds indicative of the presence of other oils.

The multiple assignment recovered analysis (MARA) on NMR spectra was used for a direct food labeling in olive oils by Rotando *et al* (2019). The method takes advantage of the multiple NMR signals generated by any chemical; these will be all proportional to the concentration of the parent compound. They show that MARA–NMR is an effective, innovative, and quick method for food labeling; unlike other analytical techniques it is self-consistent, smoothing out random instrumental outliers or unpredictable anomalies.

How the concentration of oil and water can be determined from the NMR-MOUSE signals was demonstrated by Pedersen *et al* (2003). The NMR-MOUSE is a small and portable LF-NMR system with a one-sided magnet layout that replaces the conventional magnet and probe on an LF-NMR instrument. Since the magnetic field strength rapidly decreases as a function of the distance from the surface of the NMR-MOUSE, measurements are only possible close to the surface of the sample. Pedersen *et al* (2003) has demonstrated how one could use the NMR-MOUSE to obtain quantitative measurements on model food systems.

3.2.4 Application of NMR spectroscopy to specific foods

Gall *et al* (2004) proposed the use of NMR technique to establish if teas could be discriminated according to the country of origin or with respect to quality. About 30 compounds were identified in the 1D and 2D spectra, and more than 50 signals or groups of signals were indexed overall. Sucrose, glucose, amino acids, fatty acids, phenolics, flavonoids (flavan-3-ols or catechins, flavonols), xanthines, and minor sugars have been observed at this study.

Charlton *et al* (2002) studied quality control and authenticity of instant coffee by ^{1}H NMR spectroscopy. Researchers have demonstrated, using statistical methods, the presence of inherent differences between coffees produced by different manufacturers, and even between those produced by the same manufacturer, by identifying 5-(hydroxymethyl)-2-furaldehyde as a marker compound using the structural characteristics determined by NMR.

The application of ^{1}H NMR profiling for the differentiation of organic and conventional *Coffea arabica* roasted coffee samples has been research by Consonni *et al* (2018). As a matter of fact, the wide-range metabolic profiling performed by NMR, allowed the detection of different chemical compounds simultaneously, thus providing a detailed information about the metabolite content that is influenced by the farming processes. The results of this research presented ^{1}H NMR spectroscopy as a valid method for farming differentiation of *C. arabica* roasted coffee. The ^{1}H NMR spectra of water extracts from *C. arabica* roasted coffee are characterized by the presence of major soluble metabolites, like chlorogenic acids (caffeoyl/feruloyl-quinic acids, namely CGA), trigonelline, *N*-methyl-pyridine, and caffeine dominating the aromatic region while organic acids (acetate, citrate, lactate, malate, and quinic acid), fatty acids, sucrose and other small components were present in the aliphatic region.

NMR spectroscopy in combination with multivariate data analysis is a suitable tool to screen eggs according to the different systems of husbandry. Hence, Ackermann *et al* (2019) utilized NMR technique for classification of organically produced chicken eggs in Germany and achieved a success rate of about 93%.

For honey, it is possible to verify the botanical origin and exclude adulteration with sugars by NMR. Honey is a very complex multi-component system and its LF

^{1}H NMR relaxation profile can be modeled as a linear combination of characteristic relaxation times from the measurable hydrogens present in its structure. In LF ^{1}H NMR studies, proton relaxation is described by the relaxation time constants T1 (longitudinal) and T2 (transverse), where T2 relaxation decay in food is multi-exponential, indicating the presence of different water populations or water 'pools' in the foods matrixes (Oliveira *et al* 2014). In spice analysis, NMR allows one to detect crude adulterations (e.g. of saffron) or quantify marker ingredients such as essential oils (Kuballa *et al* 2018). Also, Oliveira *et al* (2014) used NMR for classification of Brazilian honeys. They demonstrated that LF ^{1}H NMR can be a viable technique for use in classifying honeys by their botanical origin.

3.2.5 Application of NMR spectroscopy to fruit and vegetables

NMR spectroscopy has been used to investigate the determination of water content and water distribution in several foods. NMR used changes in subcellular water compartmentation in parenchyma apple tissue during drying and freezing in a study by Hills and Remigereau (1997). This research has illustrated the power of NMR relaxometry for monitoring subcellular water compartmentation. The potential of NMR and magnetic resonance imaging (MRI) for non-invasively monitoring the subcellular and intercellular redistribution of water in cellular tissue during drying and freezing processes was assessed and it was concluded that nonspatially resolved NMR relaxation and diffusion techniques still provided the best probes of subcellular water compartmentation in tissue.

Richardson *et al* (2019) have applied NMR spectroscopy in combination with chemometrics to quantify the adulteration of fresh coconut water, stretched with water–sugar mixtures. Researchers have concluded that ^{1}H NMR spectroscopy enables accurate quantification for the degree of adulteration in this product, with the added discovery that the shift and line shape of the malic acid signal can be utilised as a potential diagnostic marker for partial substitution of fresh coconut water with extrinsic components such as sugar mixtures.

3.2.6 Application of NMR spectroscopy to meat products

The difference of metabolite profiles between raw and cooked pufferfish (*Takifugu flavidus*) meat was explored by ^{1}H NMR technique by Yang *et al* (2019). All the metabolites identified using a list of 2D NMR experiments by both ^{1}H and ^{13}C data. Based on the 2D NMR analysis, 24 metabolites, including 13 amino acids (valine, isoleucine, methionine, glutamate, arginine, lysine, aspartate, leucine, glycine, tyrosine, tryptophan, alanine and taurine), 5 organic acids (succinate, lactate, acetate, creatine and fumarate), 4 nucleic acids (inosine, 5′-IMP, uracil and hypoxanthine), and 2 alkaloids (betaine and choline) have been identified.

Sacco *et al* (2010) studied meat components and concluded that this method makes it possible to acquire qualitative and quantitative information about chemical composition, both quickly and without any particular preparation of the sample to be analysed. They determined the potentiality of this method in defining the origin of meat and the possibility of identifying meat adulteration. Leffler *et al* (2008) determined fat and moisture content in chicken, pork, beef, turkey, beef hot dog and pork sausage by NMR. Results demonstrated that microwave drying with NMR is a rapid, practical method in raw and processed meat products.

3.2.7 Application of NMR spectroscopy to food emulsions

The unbiased and quantitative nature of ^{1}H NMR has been exploited to assess lipid oxidation products in mayonnaise, a particularly oxidation-prone food emulsion, by Merkx *et al* (2018). ^{1}H NMR signals of hydroperoxides have been assigned in a fatty acid and isomer specific way.

NMR has been demonstrated in a truly non-invasive through-package 'sensor' mode, also denoted as the MObile Universal Surface Explorer (MOUSE). MOUSE sensor has been used for assessment of the microstructural quality of a food material (Haiduca *et al* 2007). The NMR-MOUSE is a small and portable LF-NMR system with a one-sided magnet layout that is used to replace the conventional magnet and probe on an LF-NMR instrument. The high magnetic field gradients associated with the one-sided MOUSE magnet result in NMR signal decays being dominated by molecular diffusion effects, which makes it possible to discriminate between the NMR signals from oil and water (Pedersen *et al* 2003). They had taken model systems consisting of protein-stabilized oil-in-water emulsions, where an important microstructural quality parameter is water exudation. They found that the performance of the MOUSE is comparable to that of conventional benchtop NMR. Haiduca *et al* (2007) concluded that low field NMR relaxometry can be deployed to assess the microstructural quality of food products, by both conventional benchtop NMR as well as the NMR-MOUSE. They concluded that NMR-MOUSE is a useful technique within the area of food analysis (Haiduca *et al* 2007, Pedersen *et al* 2003).

3.3 Results

NMR spectroscopy is a wonderful tool for the compositional analysis of foods. It is a powerful technique that can provide information about samples besides the molecular structure, purity and content. It is used for determination of the quality characteristics of foods successfully. NMR is an extremely reliable analytical method that can get results in a short time combined with ease of sample preparation. The results obtained from the studies are proving that NMR technique can be used successfully in foods for determination of properties of the composition, monitoring of water mobility, monitoring of amino acids, organic acids and fatty acids, characterization of geographical origin and maturation time.

References

Ackermann S M, Lachenmeier D W, Kuballa T, Schütz B, Spraul M and Bunzel M 2019 NMR-based differentiation of conventionally from organically produced chicken eggs in Germany *Magn. Reson. Chem.* **57** 579–88

Belloque J and Ramos M 1999 Application of NMR spectroscopy to milk and dairy products *Trends Food Sci. Technol.* **10** 313–20

Belton P S *et al* 1998 Application of Chemometrics to the ^{1}H NMR spectra of apple juices: discrimination between apple varieties *Food Chem.* **61** 207–13

Brescia M A, Monfreda M, Buccolieri A and Carrino C 2005 Characterisation of the geographical origin of buffalo milk and mozzarella cheese by means of analytical and spectroscopic determinations *Food Chem.* **89** 139–47

Charlton A J, Farrngton W H H and Brereton P 2002 Application of ^{1}H NMR and multivariate statistics for screening complex mixtures: quality control and authenticity of instant coffee *J. Agric. Food Chem.* **50** 3098–103

Consonni R, Polla D and Cagliani L R 2018 Organic and conventional coffee differentiation by NMR spectroscopy *Food Control.* **94** 284–8

Curtıs S D A, Curını R, Delfını M, Brosıo E, D'ascenzo F and Bocca B 2000 Amino acid profile in the ripening of Grana Padano Cheese: a NMR study *Food Chem.* **71** 495–502

Gall G L, Colquhoun I J and Defernez M 2004 Metabolite profiling using ^{1}H NMR spectroscopy for quality assessment of green tea *Camellia sinensis* (L.) *J. Agric. Food Chem.* **52** 692–700

Gil A M and Rodrigues J 2008 Prevention methods for the characterization of beer by nuclear magnetic resonance spectroscopy *Beer in Health and Disease* (Amsterdam: Elsevier), pp 935–42

Haiduca M, Trezzaa E E, Dusschotenb D V, Reszkac A A and Duynhoven J P M V 2007 Non-invasive 'through-package' assessment of the microstructural quality of a model food emulsion by the NMR MOUSE *LWT* **40** 737–43

Hills B P and Remigereau B 1997 NMR studies of changes in subcellular water compartmentation in parenchyma apple tissue during drying and freezing *Int. J. Food Sci. Technol.* **32** 51–61

Hong Y S 2012 NMR-based metabolomics in wine science. magnetic resonance in food-dealing with complex systems *Magn. Reson. Chem.* **49** S13–21 (Special Issue: Magnetic Resonance in Food: Dealing with Complex Systems)

Kaffarnik S, Ehlers I, Gröbner G, Schleucher J and Vetter W 2013 Two-dimensional ^{31}P,^{1}H NMR spectroscopic profiling of phospholipids in cheese and fish *J. Agric. Food Chem.* **61** 7061–9

Keeler J 2010 *Understanding NMR Spectroscopy* 2nd edn (Chichester: Wiley)

Koda M, Furihata K, Wei F, Miyakawa T and Tanokura M 2012 Metabolic discrimination of mango juice from various cultivars by band-selective NMR spectroscopy *J. Agric. Food Chem.* **60** 1158–66

Kuballa T, Brunner T S, Thongpanchang T, Walch S G and Lachenmeier D W 2018 Application of NMR for authentication of honey, beer and spices *Curr. Opin. Food Sci.* **19** 57–62

Leffler T P, Moser C R, Mcmanus J and Urh J J 2008 Determination of moisture and fat in meats by microwave and nuclear magnetic resonance analysis: collaborative study *J. AOAC Int.* **91** 802–10

Mannina L, Marini F, Antiochia R, Cesa S, Magrí A, Capitani D and Sobolev A P 2010 Tracing the origin of beer samples by NMR and chemometrics: trappist beers as a case study *Electrophoresis* **37** 2710–9

Maruyama Y, Numamoto Y, Saito H, Kita R, Shinyashiki N, Yagihara S and Fukuzaki M 2014 Complementary analyses of fractal and dynamic water structures in protein–water mixtures and cheeses *Colloids Surf.* A **440** 42–8

Merkx D W H, Hong G T S, Ermacora A and Duynhoven J P M V 2018 Rapid quantitative profiling of lipid oxidation products in a food emulsion by ^{1}H NMR *Anal. Chem.* **90** 4863–70

Oliveira R D, Ribeiro R, Mársico E T, Carneiro C D S, Monteiro M L G, Júnior C A C, Mano S and Jesus E F O 2014 Classification of Brazilian honeys by physical and chemical analytical methods and low field nuclear magnetic resonance (LF 1H NMR) *LWT - Food Sci. Technol.* **55** 90–5

Parlak Y 2016 Reduction opportunities of sodium level in Circassian cheese by using salt substitute *PhD Thesis* (Adana, Turkey: Cukurova University)

Pedersen H T, Ablett S, Martin D R, Mallett M J D and Engelsen S B 2003 Application of the NMR-MOUSE to food emulsions *J. Magn. Reson.* **165** 49–58

Proton Nuclear Magnetic Resonance Spectroscopy (H-NMR) http://chem.ucla.edu/~harding/notes/notes_14C_nmr02.pdf [accessed 20 July 2019]

Richardson P I C, Muhamadali H, Lei Y, Golovanov A P, Ellis D I and Goodacre R 2019 Detection of the adulteration of fresh coconut water via NMR spectroscopy and chemometrics *Analyst* **144** 1401–8

Rotando A, Mannina L and Salvo A 2019 Multiple assignment recovered analysis (MARA) NMR for a direct food labeling: the case study of olive oils *Food Anal. Methods* **12** 1238–45

Sacco A, Vonghia G, Giannico F, Sacco D, Martino V D, Jambrenghi A C and Brescia M A 2010 High resolution nuclear magnetic resonance spectroscopy (NMR) studies on meat components: potentialities and prospects *Ital. J. Anim. Sci.* **1/2** 151–8

Schievano E, Guardini K and Mammi S 2009 Fast determination of histamine in cheese by nuclear magnetic resonance (NMR) *J. Agric. Food Chem.* **57** 2647–52

Schripsema J 2019 Similarity and differential NMR spectroscopy in metabolomics: application to the analysis of vegetable oils with ^{1}H and ^{13}C NMR *Metabolomics* **15** 39

Shintu L and Caldarelli S 2005 High-resolution MAS NMR and chemometrics: characterization of the ripening of Parmigiano Reggiano cheese *J. Agric. Food Chem.* **53** 4026–31

Silva L A, Flumignan D L, Tininis G A, Pezza H R and Pezza L 2019 Discrimination of Brazilian lager beer by ^{1}H NMR spectroscopy combined with chemometrics *Food Chem.* **272** 488–93

Tognarelli J M, Dawood M, Shariff M I F, Grover V P B, Crossey M M E, Cox I J, Taylor-Robinson S D and McPhail M J W 2015 Magnetic resonance spectroscopy: principles and techniques: lessons for clinicians *J. Clin. Exp. Hepatol.* **5** 320–8

Yang L, Dai B, Ayed C and Liu Y 2019 Comparing the metabolic profiles of raw and cooked pufferfish (*Takifugu flavidus*) meat by NMR assessment *Food Chem.* **290** 107–13

Yekta S S, Hedenström M, Svensson B H, Sundgren I, Dario M, Prast A E, Hertkorn N and Björn B 2019 Molecular characterization of particulate organic matter in full scale anaerobic digesters: an NMR spectroscopy study *Sci. Total Environ.* **685** 1107–15

Chapter 4

Spectroscopic method for the detection and determination of ammonia nitrogen in aquaculture water

Daoliang Li, Zhen Li, Cong Wang, Tan Wang and Xianbao Xu

Ammonia-N, one of the main forms of nitrogen in aquaculture water, is an indispensable source of protein synthesis in aquatic organisms. High ammonia concentration will be toxic to aquatic organisms and cause water eutrophication which threatens aquaculture production safety and destroys the ecological balance. The spectroscopic methods employed for ammonia-N detection are presented according to the following classifications: spectrophotometry, atomic absorption spectroscopy, infrared spectroscopy, fluorescence method and chromatography. In past decades, these traditional technologies have been widely used for detection of pollutants in water, but many disadvantages still exist, such as low precision, long detection time and limited research fields.

4.1 Introduction

Ammonia nitrogen, referred to as ammonia in natural waters in the form of free ammonia (NH_3) or ammonium salt (NH_4^+) (Zhang *et al* 2018), is one of the three forms of nitrogen in aquaculture water.

With the development of the economy and the expansion of the aquaculture industry (Food and Agriculture Organization 2016), ammonia nitrogen pollution in aquaculture water caused by microbial decomposition, residual bait of aquaculture, domestic sewage discharge and industrial pollution discharge all have a growing trend. Excessive ammonia in aquaculture water not only causes algae to multiply as well eutrophication, but also affects the growth and development of aquaculture objects in aquaculture water environment, and even causes a large number of deaths of fish, shrimp and crab, resulting in ecological damage and huge economic losses

doi:10.1088/2053-2563/ab4428ch4

(Cheng *et al* 2014, Kocour Kroupová *et al* 2016, Baker *et al* 2017). Furthermore, the ammonia nitrogen in drinking water can be easily transformed into nitrate (Zhou 2016), which will poison the human body. In aquaculture production, the concentration of ammonia (NH_4^+) should be controlled below 0.02 mg l^{-1}, and should not exceed 0.6 mg l^{-1} in actual aquaculture. Generally speaking, the ammonia tolerance of fish species is weaker than that of adults, and the ammonia tolerance of different fish species is different. The total ammonia content should not exceed 0.5 mg l^{-1} and the ammonia nitrogen content should be below 0.2 mg l^{-1} (Geng-Dong *et al* 2011, Fallis 2015). Therefore, there is an urgent need to detect ammonia nitrogen in aquaculture water to prevent water pollution and ensure the safety of aquatic products.

Many methods have been developed for trace level detection of ammonia nitrogen in aquaculture water. Electrochemical methodologies, including voltammetric (Ji *et al* 2010), potentiometric (Lin *et al* 2014) and impedimetric electrodes (Karthickkannan and Saraswathi 2013), convert NH_3 or NH_4^+ to current signal, potential difference and impedance, respectively. These methods are easily performed, consuming no or few reagents and requiring no complex or time-consuming pretreatment; in addition, the detection equipment is inexpensive and easily designed. Spectroscopic methodologies, including fluorescence spectrometry (Yan *et al* 2015) and absorption spectrometry (Cho *et al* 2018, Liang *et al* 2016), convert the presence of nitrite ion to optical signals. Spectroscopic methodologies can usually reach a very low detection limit with good precision. Combined with enrichment and separation methods, such as capillary electrophoresis, chromatography and liquid extraction, the detection limit can be further reduced.

Spectroscopic methods for ammonia nitrogen detection operate generally by measuring the radiation or absorption intensity of a particular wavelength affected by ammonia nitrogen. Spectroscopy is a detection method that can be incorporated with other separation and enrichment methods, such as capillary electrophoresis, chromatography and liquid–liquid extraction (Yao *et al* 2012, Chen *et al* 2011), to improve detection accuracy and decrease the detection limit. A number of excellent reviews have been reported over the past decade (Valente *et al* 2017, He *et al* 2019). However, detection requirements and technology have developed in recent years. The aim of this article is to review various spectroscopic methods for ammonia nitrogen detection in aquaculture water reported in recent years and to summarize their advantages and disadvantages. It is organized based on the ways they are produced, including absorption spectrometry and emission spectrometry. Moreover, this article provides a broad summary of research findings that use different spectroscopic methods for detection of ammonia nitrogen in an aqueous environment together with their characteristic properties. Finally, the advantages/disadvantages and limitations of each technique are discussed. It is intended as a guide for further studies that concentrate on detection of other contaminants using Raman spectroscopy, particularly ammonia nitrogen.

4.2 Spectroscopic method for the detection of ammonia nitrogen

4.2.1 Absorption spectrometry

Absorption spectrum is a common method in the detection of ammonia nitrogen in water, especially electronic colorimetry and UV–vis absorption spectroscopy. Table 4.1 lists some research results in recent years.

4.2.1.1 Colorimetric spectrophotometry

The colorimetric method (spectrophotometry) is a method for qualitative analysis using the absorption characteristics of a specific substance of light by a colored substance, and the principle is based on the color of the solution of the substance to be tested or the color of the colored solution formed by adding the color developer. The color depth is proportional to the amount of matter, and the amount of material in the solution can be determined based on the intensity of light absorbed by the colored solution. Based on the color reaction of the colored compound, the basic requirement of colorimetric reaction for color reaction is that the reaction should have high sensitivity and selectivity, and the composition of the colored compound formed by the reaction is constant and stable, and it develops color. The color of the agent varies greatly. Choosing the appropriate color reaction and controlling the appropriate reaction conditions is the key to colorimetric analysis.

The colorimetric method is the earlier detection method in ammonia nitrogen measurement. Traditional color reagents are represented by Nessler reagent. At present, Nessler colorimetry has become the standard test method in some countries. However, the Nessler colorimetric method requires the use of the highly toxic substance mercury iodide. Therefore, the current color development reagents gradually adopt non-toxic and low-pollution indophenol blue coloring agents such as salicylic acid and o-phenylphenol (Ma *et al* 2017). In recent years, colorimetry is mainly photoelectric colorimetry, also known as spectrophotometry. Its research is mainly reflected in the wide application of mobile technology. Zhu *et al* (2014) used flow injection technology combined with 2.5 m liquid waveguide capillary to establish on-line determination of seawater. The automatic colorimetric method of trace ammonium has high sensitivity and the detection limit is 3.6 nmol L^{-1}, the linearity is 10–500 nmol L^{-1}, but the detection upper limit of 30 μmol L^{-1} can be achieved by selecting an insensitive detection wavelength or a lower reaction temperature. In addition to the on-line detection, the method is more convenient for detection because it reduces reagent consumption, high sample throughput. Cogan *et al* (2014) have combined microfluidic technology with LED-based optical inspection systems to develop a low-cost monitoring system for detecting ammonia in freshwater and wastewater. With the robustness and low cost of microfluidic platforms and integrated wireless communications, it has become the ideal level for on-site environmental monitoring. For online detection, Lin *et al* (2018) used reverse flow injection analysis (rFIA) to achieve ammonia nitrogen detection with detection limits of 0.07 μM (fresh water) and 0.08 μM (seawater). Kodama *et al* (2015) used flow analysis combined with a 1 m long liquid capillary tube and fiber spectrometer

Table 4.1. Parameters and performances of absorption spectrometry.

Detection	Material	Emission	Detection range	LOD	RSD	Ref.
Colorimetric	Indophenol blue	690 nm	10–500 nmol L^{-1}	3.6 nmol L^{-1}	4.4%	Zhu *et al* (2014)
Colorimetric	Berthelot reagent	660 nm	12 mg l^{-1}	0.015 mg l^{-1}	N/A	Cogan *et al* (2014)
Colorimetric	Indophenol blue with o-phenylphenol	690 nm	50 μmol L^{-1}, 35 μmol L^{-1}	0.07 μmol L^{-1}, 0.08 μmol L^{-1}	<1.3%	Lin *et al* (2018)
Colorimetric	Indophenol blue	630 nm, 530 nm	2000 nM, 10 000 nM	5.5 ± 1.8 nM, 13 ± 5.3 nM	N/A	Kodama *et al* (2015)
UV–vis	N/A	460 nm, 515 nm	5–100 ppm	N/A	N/A	Dubas and Pimpan (2008)
UV–vis	Ag $(NH_3)_2^+$	400 nm	10–1000 mg l^{-1}	180 mg l^{-1}	N/A	Amirjani and Fatmehsari (2018)
UV–vis		N/A	N/A	1 ppm		Pandey *et al* (2012)
UV–vis	$OHDBMBF_2$	500 nm	N/A		N/A	Gelfand *et al* (2018)
Infrared	aza-BODIPYs dyes	N/A	300 μg L^{-1}	0.11 μg L^{-1}	N/A	Strobl *et al* (2017)
Chromatography	Bromothymol blue	625 nm	0.05–0.26, 0.26–2.62 mg l^{-1}	0.05 mg l^{-1}	⩽1.0%	Yao *et al* (2012)
Gas-phase molecular absorption spectrometry	Hypobromous acid	N/A	0.10–2.00 mg l^{-1}	0.003 mg l^{-1}	1.0%–1.6%	Liu *et al* (2016)

to record multi-wavelength absorbance. Due to the convenient blank comparison, it was suitable for the determination of ammonium ion concentration in the ocean.

4.2.1.2 UV–vis absorption spectroscopy (UV–vis)

UV–vis absorption spectroscopy is an analytical method based on the selective absorption of ultraviolet–visible light by molecules of a test substance. This molecular absorption spectrum is generated between electrons in valence electrons and molecular orbitals. Transition is widely used for qualitative and quantitative determination of organic and inorganic substances. The method has the characteristics of high sensitivity, good accuracy, excellent selectivity, easy operation and good analysis speed.

The characteristics of ammonia-sensitive particles determine the requirements of the detection spectrum. It is reported that the position and amplitude of the silver-positioned plasmon resonance band are monitored by a UV–vis spectrometer to detect ammonia in water. Dubas and Pimpan (2008) monitored silver ions under weak ultraviolet irradiation. The reduction was used to detect ammonia. Similarly, Amirjani and Fatmehsari (2018) manipulated the surface plasma bands of silver nanoparticles (AgNPs) by the formation of silver $(NH_3)^{2+}$ complexes. This complex reduces the amount of AgNPs in the solution, thereby reducing the color strength of the colloidal system. The color intensity change of the solution was followed by using a UV–vis spectrophotometer to achieve a detection limit of 180 mg l^{-1}. In addition, Pandey *et al* (2012) developed an ammonia optical sensor based on a green synthetic biopolymer–nanosilver nanocomposite. Gelfand *et al* (2018) studied the spectral characteristics of a novel organoboron dye o-hydroxydibenzoylmethane boron difluoride ($OHDBMBF_2$) using UV–vis absorption spectroscopy and quantum chemistry.

4.2.1.3 Infrared spectroscopy

Infrared spectroscopy is a method for quantitative and qualitative analysis of various infrared absorbing compounds based on the selective absorption of electromagnetic radiation from different areas of the infrared light. A beam of infrared light of different wavelengths is irradiated onto the molecules of the substance, and certain specific wavelengths of infrared rays are absorbed to form an infrared absorption spectrum of the molecule.

The infrared spectrum of ammonia nitrogen detection in water is mainly in the near-infrared band. Due to the absorption of infrared light by water, it is very difficult to detect the ammonia nitrogen in water by infrared, which puts high requirements on sensitive reagents. Strobl *et al* (2017) based on fluorescent BF_2-chelated tetraarylazadipyrromethene dyes (aza-BODIPYs) dyes, using near-infrared (NIR)-emitting Egyptian blue as a reference material sensor, combined with optical fiber to achieve ammonia detection in water, the most sensitive sensor has a detection limit of 0.11 μg L^{-1} and an upper detectable concentration of 300 μg L^{-1}.

4.2.1.4 Other methods
Yao *et al* (2012) designed a small-size gas-tight optical measuring system for detection of ammonia nitrogen in water, prepared based on gas-phase ammonia induced color change of the sensing element that was made by loading bromothymol blue (BTB) in a transparent porous glass fiber membrane. A 625 nm light emitting diode (LED), a photodetector and a sensing element were mounted in the gas-testing chamber for optical response to ammonia gas released from the water in the liquid-sample chamber. Owing to the amount accumulation of ammonia gas in the sealing system, the ammonia nitrogen detection limit of the device can research 0.05 mg l^{-1}. Two linear-response ranges from 0.05 mg l^{-1} to 0.26 mg l^{-1} and from 0.26 mg l^{-1} to 2.62 mg l^{-1} were obtained.

Liu *et al* (2016) used the gas-phase molecular absorption spectrometry method employed to determine the content of ammonia nitrogen in seawater. The results showed that good linear relationships were obtained for ammonia nitrogen in the mass concentration ranges of 0.01–0.40 mg l^{-1} and 0.10–2.00 mg l^{-1}, respectively, and the coefficient values were both greater than 0.999. The method detection limit was 0.003 mg l^{-1}. RSDs of the actual samples and standards were in the range of 1.0% –1.6%, and the recovery rates were in the range of 94.0%–110%.

4.3 Fluorescence spectrometry

Fluorescence spectroscopy refers to the process in which a specific wavelength of light is irradiated into a solution, and the fluorescent substance in the solution absorbs energy-releasing energy. The molecule absorbs light, transitions from the ground state to the unstable excited state, and then returns from the unstable excited state to the ground state. When the excited state electrons re-transition from the lowest vibrational energy level to the different vibrational levels of the ground state, energy is emitted in the form of fluorescence. Since the energy level difference between the excited state and the ground state of different elements is different, the wavelengths of fluorescence generated are different. Quantitative and qualitative analysis of substances in samples is based on the wavelength and intensity of the fluorescent light (Xu and Wang 2006). Fluorescence spectroscopy is widely used to detect ammonia nitrogen in water. This chapter mainly summarizes various methods to detect ammonia nitrogen in water. In addition, we summarize the parameters and results of these methods in table 4.2.

4.3.1 O-phthalaldehyde (OPA)

The principle of fluorescence measurement of ammonia nitrogen is based on the fluorescence derivatization reaction between o-phthalaldehyde (OPA) and ammonia nitrogen. The fluorescence intensity and concentration of the resulting derivative conform to Lambert–Beer law within a certain range. Fluorescence intensity can be used to extract the ammonia nitrogen concentration of the sample. In 1971, Roth (1971) discovered through experiments that a sensitive fluorescence reaction occurs between ammonia nitrogen and OPA and mercaptoethanol, that is, the fluorescence isomerization substitution derivatization reaction. In the following decades, many

Table 4.2. Parameters and performances of fluorescence spectrometry.

Method	Material	Exciting/Emission	Detection range	LOD	RSD	Ref.
Fluorescence	OPA–mercaptoethanol	410/470 nm	0.002–10 mM	N/A	N/A	Goyal *et al* (1988)
Fluorescence	OPA	365/425 nm	1–1000 μM	20 nM	N/A	Zhang and Dasgupta (1989)
Fluorescence	OPA, 2-mercaptoethanol	415/486 nm	0–0.7 mg l^{-1}	0.0012 mg l^{-1}	1.3%	Xie *et al* (1999)
Fluorescence	OPA	360/425 nm	0–10 μM L^{-1}	0.0025 M L^{-1}	<1%	Xiang-xiang and Wei-dong (2007)
Fluorescence	OPA	363/426 nm	2–300 μg L^{-1}	1.95 μg L^{-1}	<3%	Wang *et al* (2010)
Fluorescence	OPA–sulfite	365/430 nm	0–419.31 μg L^{-1}	0.16 μg L^{-1}	N/A	Wang *et al* (2018)
Fluorescence	Oxazine 170 perchlorate	N/A/565,630 nm	1–60 ppm	N/A	N/A	Duong and Rhee (2014)
Fluorescence	4-Methoxyphthalaldehyde	370/454 nm	0.025–0.3 μmol L^{-1}	0.0058 μmol L^{-1}	N/A	Liang *et al* (2015)
Fluorescence	4,5-Dimethoxyphthalaldehyde, Sulfite	380/485 nm	0–5 μM,0–2 μM	6.5 nM,3.5 nM	N/A	Zhang *et al* (2018)

scholars have studied the derivatization reaction to measure ammonia nitrogen in seawater. In recent years, many optimizations have been made in this method.

Goyal *et al* (1988) used OPA–mercaptoethanol mixed reagent to react with ammonia nitrogen in water to form a derivative with strong fluorescence. The excitation and emission peaks of this derivative are 410 nm and 470 nm, respectively, the linear range was 0.002–10 mM. This method of detecting ammonia nitrogen has a high selectivity, but the disadvantage is that the method has a great influence on the seawater matrix effect.

Genfa *et al* (Zhang and Dasgupta 1989) proposed a method for the detection of ammonium ions based on OPA-flow injection fluorescence analysis. It was found through experiments that the fluorescence intensity of ammonia nitrogen was little affected by the ionic strength of the system. In the experiment, the optimum excitation/emission wavelength is 365 nm/425 nm, the linear range was 1–1000 μM and LOD was 20 nM, sodium sulfite was substituted for mercaptoethanol reagent to participate in the reaction, and it was found that sodium sulfite stabilized well. Holmes *et al* (Kérouel and Aminot 1997, Aminot *et al* 2001, Holmes *et al* 1999) improved the method based on Genfa's research, making the method more suitable for the analysis of ammonia nitrogen in sea areas with high salinity and estuary.

Xie *et al* (1999) have found a flow injection fluorescence analysis method based on OPA, 2-mercaptoethanol (reducing agent) and ammonia nitrogen in alkaline conditions. The research focuses on the spectral characteristics, pH and other factors affecting the determination of ammonia (ammonium ion). The linear range of ammonia (ammonium ion) was 0–0.7 mg L^{-1}, the relative standard deviation was 1.3% (CNH_3 = 0.2 mg L^{-1}, n = 11), and the detection limit was 0.0012 mg L^{-1}. Based on the reaction system, Guo *et al* (2004) established an analytical method for the determination of trace ammonia nitrogen in water by direct and flow injection fluorimetry. The experimental study was carried out on the pH of the reaction system and the addition of nonionic surfactant Triton X-100. At λ_{ex} = 415 nm, λ_{em} = 486 nm, the direct fluorescence method was used to determine the NH_3 content in the range of 0.2–1.0 μg ml^{-1}, and the linear range determined by the flow injection method was 0–0.7 μg ml^{-1} and the linear correlation coefficient R of the working curve can reach 0.9995 and 0.9998.

Yu and Guo (2007) studied the fluorescence spectroscopy characteristics on the basis of Holmes, and compared several commonly used ammonia nitrogen measurement methods, and proposed an improved fluorescence analysis method. The results show that the improved method is suitable for the measurement of low ammonia nitrogen in seawater. It has been found that dissolved organic matter, Ca^{2+}, Mg^{2+}, etc in seawater compete with ammonia nitrogen for OPA, thereby reducing the amount of ammonia nitrogen–OPA derivatives produced in seawater samples, and the uncorrected recovery rate is reduced by 3%–7%. The improved method has a linear range of 0–10 μmol L^{-1} and a detection limit of 0.0025 mol L^{-1}.

Wang *et al* (2010) analyzed the experimental conditions of the fluorescence derivatization of ammonia and OPA. The linear relationship between the fluorescence intensity and the presence of ammonia nitrogen in the water sample determined the content of ammonia nitrogen in seawater. The experiment did not require heating. At the excitation/emission wavelength of 363 nm/426 nm, the linear range of detection is 2–300 μg L^{-1}, the RSD is less than 3%, and the method detection concentration is 1.95 μg L^{-1}. The results show that the pH of the system has an effect on the fluorescence intensity.

Wang *et al* (2018) studied the ammonia–phthalic acid diester–sulfite complexation reaction in alkaline aqueous solution to study the change of fluorescence intensity under different pH, temperature, salinity and atmospheric pressure conditions, and developed it in natural water. The ammonia nitrogen sensor used, for the first time, introduced random sequence modulation and self-calibration techniques into fluorescence amplitude measurements. The sensor detection range is 0–419.31 μg L^{-1}, the detection limit is 0.16 μg L^{-1}, and the R^2 is 0.99.

4.3.2 Other methods

There are also some fluorescent methods added to other reagents, such as oxazine 170 perchlorate (O17), 4-methoxyphthalaldehyde (MOPA). Duong and Rhee (2014) developed a fluorescence proportional ammonia sensor using the change of fluorescence emission spectrum when prochlorazide 170 (O17) was used to remove protons in water and combined with ammonium ions. Ratio fluorescence sensing was performed using a ratio of fluorescence intensities at two emission wavelengths, EM = 565 nm and EM = 630 nm. O17 has good compatibility with EC, and has a short response time (t_{95} = 10 s), good flammability, low interference, long-term stability, and detection range of 1–60 ppm. The ammonia content in the artificial wastewater containing different ions was determined by ammonia sensing membrane (O17–EC membrane), and the obtained results were in good agreement with the ammonia concentration in the artificial wastewater samples.

Liang *et al* (2015) successfully synthesized 4-methoxyphthalaldehyde (MOPA), a new fluorescent reagent for the determination of ammonium. Using the excitation wavelength of 370 nm and the emission wavelength of 454 nm, a new fluorescence analysis method was established to detect trace amounts of ammonium in natural water. The results showed that the optimized MOPA concentration was 0.12 g L^{-1}, the pH range was 11.2–12.0, and the sulfite concentration was 0.051 g l^{-1}. The reaction time was 15 min, the linear range of the method was 0.025–0.300 μmol L^{-1}, and the method detection limit was 0.0058 μmol L^{-1}. The matrix recovery of the method was 93.6%–108.1%. Compared with the OPA method, this method is sensitive, fast, and free from background peaks, and is more suitable for developing portable fluorescence detection systems.

4,5-Dimethoxyphthalaldehyde (M_2OPA) was also used to determine ammonium in ambient water (Zhang *et al* 2018). Ammonium reacts with M_2OPA and sulfite to form a fluorescent derivative with maximum excitation (λ_{ex}) and emission (λ_{em}) at 380 and 485 nm, while OMA reacts with λ_{ex} and λ_{em} of 360 and 430 nm. The M_2OPA assay has a fast response and enhanced fluorescence intensity compared to the OPA based assay. The linear detection range of the single laser beam and the dual laser beam is 0–5 μM and 0–2 μM, respectively, and the detection limit is 6.5 nM and 3.5 nM.

4.4 Conclusion and future perspectives

The detection of ammonia nitrogen in aquaculture water is the precondition of realizing aquaculture water quality monitoring and aquaculture water safety control. The domestic and foreign research investment and application demand in this field will increase greatly. At present, there are many detection methods for ammonia nitrogen, and several existing detection methods have their own characteristics. At present, the most widely used methods are electrochemical method and spectrophotometry. Electrochemical sensors are suitable for miniaturization and long-term monitoring. Compared with spectroscopy, their detection limit is somewhat higher. But they are easily used and require no reagents or complex instruments. Spectroscopic methodologies can get very low detection limits and can be used to detect trace amounts. At the same time, reagents are required by spectroscopic methods to perform detection. Reagent consumption has been observably reduced by microfluidic systems.

The search for an accurate, fast and sensitive ammonia nitrogen on-line detection sensor is the focus of future research. In recent years, with the continuous development of sample pretreatment technology, the detection sensitivity of traditional spectrophotometry has been greatly improved because of the use of gas diffusion separation, liquid–liquid extraction, and solid phase extraction. Solid phase extraction (SPE) technology has the characteristics of high separation and enrichment ratio, low sample consumption, simple and fast operation and on-line extraction. The limit of detection of ammonia nitrogen content in seawater based on SPE technology and spectrophotometry has reached the nanomolar level. In addition, the combination of flow injection analysis and spectrophotometry can solve the problem. Moreover, the combination of flow injection analysis and spectrophotometry can solve the problem of automatic detection, which can improve the detection accuracy, and is suitable for online and real-time detection of ammonia nitrogen in aquaculture water. Further work based on Raman spectroscopy, carbon dots (C-dots), and new ammonia nitrogen fluorescence probes will lead to new breakthroughs in rapid, on-line and quantitative detection of ammonia nitrogen in aquaculture water.

References

Aminot A, Kérouel R and Birot D 2001 A flow injection-fluorometric method for the determination of ammonium in fresh and saline waters with a view to *in situ* analyses *Water Res.* **35** 1777–85

Amirjani A and Fatmehsari D H 2018 Colorimetric detection of ammonia using smartphones based on localized surface plasmon resonance of silver nanoparticles *Talanta* **176** 242–6

Baker J A, Gilron G and Chalmers B A *et al* 2017 Evaluation of the effect of water type on the toxicity of nitrate to aquatic organisms *Chemosphere* **168** 435–40

Chen G, Zhang M and Zhang Z *et al* 2011 On-line solid phase extraction and spectrophotometric detection with flow technique for the determination of nanomolar level ammonium in seawater samples *Anal. Lett.* **44** 310–26

Cheng X, Zeng Y and Guo Z *et al* 2014 Diffusion of nitrogen and phosphorus across the, sediment-water interface and in seawater at aquaculture areas of Daya Bay, China *Int. J. Environ. Res. Public Health* **11** 1557–72

Cho Y B, Jeong S H, Chun H and Kim Y S 2018 Selective colorimetric detection of dissolved ammonia in water via modified Berthelot's reaction on porous paper *Sens. Actuators B Chem.* **256** 165–75

Cogan D, Cleary J and Fay C *et al* 2014 The development of an autonomous sensing platform for the monitoring of ammonia in water using a simplified Berthelot method *Anal. Methods* **6** 7606–14

Dubas S T and Pimpan V 2008 Green synthesis of silver nanoparticles for ammonia sensing *Talanta* **76** 29–33

Duong H D and Rhee J I 2014 A ratiometric fluorescence sensor for the detection of ammonia in water *Sens. Actuators* B **190** 768–74

Fallis A G 2015 Investigations of ammonia nitrogen in aquaculture: the methodology, concentrations, removal, and pond fertilization *PhD Thesis* Auburn University

Food and Agriculture Organization 2016 *The State of World Fisheries and Aquaculture*

Gelfand N A, Fedorenko E V and Mirochnik A G *et al* 2018 Interaction of hydroxy substituted dibenzoylmethanatoboron difluoride with hydrated ammonia in solution: A combined spectroscopic and computational study *J. Mol. Struct.* **1175** 601–8

Geng-Dong H U, Chao S and Chen J Z *et al* 2011 Modeling of water circulating pond aquaculture system and its N & P removal effect *J. Ecol. Rural Environ.* **23** 857–63

Goyal S S, Rains D W and Huffaker R C 1988 Determination of ammonium ion by fluorometry or spectrophotometry after on-line derivatization with o-phthalaldehyde *Anal. Chem.* **60** 175–9

Guo L-Q, Xie Z-H and Lin X-C *et al* 2004 Direct fluorophotometric and flow-injection fluorophotometric methods for the determination of trace ammonia *Spectrosc. Spectral Anal.* **24** 851–4

He J, Yan X, Liu A, You R, Liu F and Li S *et al* 2019 The rapid-response room temperature planar type gas sensor based on the DPA-Ph-DBPzDCN for sensitive detection of NH_3 *J. Mater. Chem.* A **7** 4744–50

Holmes R M, Aminot A and Kérouel R *et al* 1999 A simple and precise method for measuring ammonium in marine and freshwater ecosystems *Can. J. Fish. Aquat. Sci.* **56** 1801–8

Ji X, Banks C, Silvester D, Aldous L, Hardacre C and Compton R 2010 Electrochemical ammonia gas sensing in nonaqueous systems: a comparison of propylene carbonate with room temperature ionic liquids *Electroanalysis* **19** 2194–201

Karthickkanna P and Saraswathi R 2013 An impedimetric ammonia sensor based on nano-structured a-Fe_2O_3 *J. Mater. Chem.* A **2** 394–401

Kérouel R and Aminot A 1997 Fluorometric determination of ammonia in sea and estuarine waters by direct segmented flow analysis *Mar. Chem.* **57** 265–75

Kocour Kroupová H, Valentová O and Svobodová Z *et al* 2016 Toxic effects of nitrite on freshwater organisms: a review *Rev. Aquacult.* **10** 525–42

Kodama T, Ichikawa T and Hidaka K *et al* 2015 A highly sensitive and large concentration range colorimetric continuous flow analysis for ammonium concentration *J. Oceanogr.* **71** 65–75

Liang Y, Pan Y and Guo Q *et al* 2015 A novel analytical method for trace ammonium in freshwater and seawater using 4-methoxyphthalaldehyde as fluorescent reagent *J. Anal. Methods Chem.* **2015** 1–7

Liang Y, Yan C, Guo Q, Xu J and Hu H 2016 Spectrophotometric determination of ammonia nitrogen in water by flow injection analysis based on NH_3-o-phthalalde-hyde-Na_2SO_3, reaction *Anal. Chem. Res.* **10** 1–8

Lin K, Li P and Wu Q *et al* 2018 Automated determination of ammonium in natural waters with reverse flow injection analysis based on the indophenol blue method with, o-phenylphenol *Microchem. J.* **138** 519–25

Lin Y M, Li L Y, Hu J W, Huang X F, Zhou C and Jia M *et al* 2014 Photometric determination of ammonia nitrogen in slaughterhouse wastewater with Nessler's reagent: effects of different pretreatment methods *Adv. Mater. Res.* **955–959** 1241–4

Liu L, Tang C and Hu X *et al* 2016 Determination of ammonia nitrogen in seawater by gas-phase molecular absorption spectrometry *Environ. Monitor. Early Warn.* **8** 16–9 (in Chinese with English abstract)

Ma J, Li P and Lin K *et al* 2017 Optimization of a salinity-interference-free indophenol method for the determination of ammonium in natural waters using, o-phenylphenol *Talanta* **179** 608–14

Min Z, Tao Z and Ying L *et al* 2018 Toward sensitive determination of ammonium in field: A novel fluorescent probe, 4,5-dimethoxyphthalaldehyde along with a hand-held portable laser diode fluorometer *Sens. Actuators* B **276** 1467–84

Pandey S, Goswami G K and Nanda K K 2012 Green synthesis of biopolymer–silver nanoparticle nanocomposite: An optical sensor for ammonia detection *Int. J. Biol. Macromol.* **51** 583–9

Roth M 1971 Fluorescence reaction for amino acids *Anal. Chem.* **43** 880–2

Strobl M, Walcher A and Mayr T *et al* 2017 Trace ammonia sensors based on fluorescent near-infrared-emitting aza-BODIPY dye *Anal. Chem.* **89** 2859–65

Valente I M, Oliveira H M, Vaz C D, Rui M R, Fonseca A J M and Cabrita A R J 2017 Determination of ammonia nitrogen in solid and liquid high-complex matrices using one-step gas-diffusion microextraction and fluorimetric detection *Talanta* **167** 747–53

Wang N, Wang C and Ha Q *et al* 2010 Fluorometric method for the determination of ammonium in seawater *Ocean Technol.* **29** 20–2

Wang C, Li Z and Pan Z *et al* 2018 Development and characterization of a highly sensitive fluorometric transducer for ultra-low aqueous ammonia nitrogen measurements in aquaculture *Comput. Electron. Agric.* **150** 364–73

Xu J and Wang Z 2006 *Fluorescence Analysis* 3rd edn (Beijing: Science press), pp 3–16 64–9

Xie Z-H, Ke Z-H and Guo L *et al* 1999 Determination of trace ammonia in water by fluorescence spectrophotometry combining with flow-injection analysis *J. Fuzhou Univ. (Nat. Sci.)* **27** 184–5

Yan C M, Liu X, Liang Y and Yin S M 2015 Determination of ammonia nitrogen in natural water sample using solid-phase fluorescence spectrometry based on OPA-NH_3-sulfite reaction *J. Instrum. Anal.* **34** 1271–5

Yao Y Q, Lu D-F and Qi Z-M *et al* 2012 Miniaturized optical system for detection of ammonia nitrogen in water based on gas-phase colorimetry *Anal. Lett.* **45** 2176–84

Yu X-X and Guo W-D 2007 Sensitive spectrofluorimetric method for determination of low concentration ammonium in seawater *Ocean Technol.* **31** 37–41

Zhang G and Dasgupta P K 1989 Fluorometric measurement of aqueous ammonium ion in a flow injection system *Anal. Chem.* **61** 408–12

Zhang L, Xu E G, Li Y, Liu H, Vidal-Dorsch D E and Giesy J P 2018 Ecological risks posed by ammonia nitrogen (AN) and un-ionized ammonia (NH_3) in seven major river systems of China *Chemosphere* **202** 136–44

Zhou L 2016 Comparison of nessler, phenate, salicylate and ion selective electrode procedures for determination of total ammonia nitrogen in aquaculture *Aquaculture* **450** 187–93

Zhu Y, Yuan D and Huang Y *et al* 2014 A modified method for on-line determination of trace ammonium in seawater with a long-path liquid waveguide capillary cell and spectrophotometric detection *Mar. Chem.* **162** 114–21

Lightning Source UK Ltd.
Milton Keynes UK
UKHW050759051219
354749UK00003B/63/P